Pergamon Titles of Related Interest

Constans   MARINE SOURCES OF ENERGY
Goodman/Love   GEOTHERMAL ENERGY PROJECTS
Grenon   THE NUCLEAR APPLE AND THE SOLAR ORANGE
Murphy   ENERGY AND ENVIRONMENTAL BALANCE
Neff   THE SOCIAL COSTS OF SOLAR ENERGY
Secretariat for Future Studies   SOLAR VERSUS NUCLEAR
Ural   ENERGY RESOURCES AND CONSERVATION RELATED TO
      BUILT ENVIRONMENT

Related Journals*

ANNALS OF NUCLEAR ENERGY
ENERGY, The International Journal
GEOTHERMICS
INTERNATIONAL JOURNAL OF HYDROGEN ENERGY
NUCLEAR AND CHEMICAL WASTE MANAGEMENT
PROGRESS IN NUCLEAR ENERGY
SOLAR ENERGY
SPACE SOLAR POWER REVIEW
SUNWORLD

*Free specimen copies available upon request.

PERGAMON POLICY STUDIES
ON SCIENCE AND TECHNOLOGY

Transitional Energy Policy 1980-2030

Alternative Nuclear Technologies

Hugh B. Stewart

Pergamon Press

NEW YORK • OXFORD • TORONTO • SYDNEY • PARIS • FRANKFURT

Pergamon Press Offices:

U.S.A.   Pergamon Press Inc., Maxwell House, Fairview Park,
               Elmsford, New York 10523, U.S.A.

U.K.   Pergamon Press Ltd., Headington Hill Hall,
               Oxford OX3 0BW, England

CANADA   Pergamon of Canada, Ltd., Suite 104, 150 Consumers Road,
               Willowdale, Ontario M2J 1P9, Canada

AUSTRALIA   Pergamon Press (Aust.) Pty. Ltd., P.O. Box 544,
               Potts Point, NSW 2011, Australia

FRANCE   Pergamon Press SARL, 24 rue des Ecoles,
               75240 Paris, Cedex 05, France

FEDERAL REPUBLIC   Pergamon Press GmbH, Hammerweg 6, Postfach 1305,
OF GERMANY   6242 Kronberg/Taunus, Federal Republic of Germany

Library of Congress Cataloging in Publication Data

Stewart, Hugh B   1916-
  Transitional energy policy, 1980-2030.

  (Pergamon policy studies on science and
technology)
  Bibliography: p.
  Includes index.
  1. Energy policy—United States. 2. Atomic
energy policy—United States. 3. Nuclear
engineering—United States. I. Title. II. Series.
HD9502.U52S824   1981   333.79'0973   80-24513
ISBN 0-08-027183-9
ISBN 0-08-027182-0 (pbk.)

Printed in the United States of America

CONTENTS

PREFACE

History may mark the 1970-1980 decade as the period when world society recognized the true gravity of an impending international energy supply/demand crisis. The cost of that belated recognition, however, has already been political upheavals in the near-east, deteriorating international stability, severe inflationary pressures and the more mundane aggravation of waiting-lines at gasoline pumps. At best, the impact of the energy supply/demand crisis in the 1980s and 1990s will be measured by economic disruptions in world societies; at worst, by a disastrous nuclear holocaust. An irony associated with the spectrum of possible outcomes is the present-day preoccupation of policy planners with the suppression of nuclear technology evolution to avoid the possibility of nuclear-weapons proliferation, when a more aggressive and thoughtful guidance of nuclear-energy directions might otherwise contribute to an earlier resolution of the international and national energy problems.

In suggesting nuclear energy as a potential solution to world energy problems, though, at least three cautions are necessary, viz:

- the length of time required for the substantial deployment of any new technology may be too great relative to the time available;
- nuclear energy may not be an appropriate solution to the pressing problem of oil supply; and
- even assuming appropriate nuclear directions can be found, the institutional problems of prompt development and deployment may preclude its usefulness.

Each of these reservations is critically important. It is an intent of this book to examine all of them.

As a basis for evaluating time constraints, the first chapter of this book will address the important subject of energy growth. It is becoming generally recognized that energy conservation is our best hope for gaining time in the emerging energy-resource supply/demand crisis. But, even if energy conservation is achievable, it will undoubtedly offer only a respite for the development of a more substantial solution. A great danger is the complacency that might come from an apparent near-zero energy growth. It is highly improbable that a near-zero growth can be sustained in the U.S. for more than one or two decades. More importantly, though, it is almost certain that energy growth cannot be throttled in the rest of the world where the energy budget is already far below that of the U.S.

It is particularly important to identify the nature of the energy-resource supply/demand problem, and to examine carefully the capability of nuclear energy to provide a resolution. That problem is addressed in Chapters II and III. Since the problems are somewhat different for the U.S. and the rest of the world, attention is given to both. Resolutions to the nuclear technology problems are examined in Chapters IV and V. And resolutions to the institutional problems are examined in Chapters V and VI.

An important theme of the book is that the nuclear energy issues, as they were viewed in the period 1965-1975, have changed dramatically in the last five years. Moreover, further changes might be expected in the next ten, twenty and even thirty years. Nuclear policies and development plans have been slow in responding to the changing issues of the last few years and may again be less than responsive in the next few decades, if the pertinent issues are not correctly defined. Hence, most of the attention in this book will be directed toward the critically important "transitional period"--the period between the existing energy technologies and the ultimate self-sufficient energy systems that might be expected in 50 to 75 years.

Some attempt has been made to introduce new approaches and controversial concepts throughout this book. The use of a cycle-adjusted-logistic growth curve for projecting domestic and world energy consumption is one such concept. Appendix A attempts to put this methodology in perspective with other more traditional projection methodologies. As will be observed from data presented in Chapter I, there appears to be persuasive reasons for believing that energy growth is likely to evolve in surges or cycles rather than monotonically. While much of the energy-growth and resource-requirement data in Chapters II and III use this growth pattern for illustrative purposes, that concept is not essential to the technology and institutional conclusions of this book. The concept is, however, useful for illustrating how some significant degree of energy conservation might be realized in the next 10 to 20 years, yet a substantial energy growth could subsequently occur.

Another controversial subject might be that of the thorium fuel cycle and the role of fast-spectrum reactors during the next 25 to 50 years. Yet there is good reason to believe the rapid commercialization of the LMFBR, simply as an alternative power plant could become an enormously expensive and risky business venture without some significant change of direction. The proposed strategy modification, discussed in Chapter V, could allow a fewer number of fast-spectrum reactors to contribute a greater leverage on fuel-resource utilization within the context of a more favorable commercialization climate.

The argument can be made, of course, that the commercialization of the thorium cycle could also be expensive. However, this latter expense would probably be an order of magnitude less than that for introducing the LMFBR in the traditionally-proposed role. Moreover, an earlier commercialization of the thorium cycle could encourage the development and deployment of more efficient thermal-spectrum reactors, a goal that should be beneficial for long-range planning. Perhaps most importantly, the combination of a fast-spectrum transmuter reactor, a thermal spectrum near-breeder reactor, the thorium fuel cycle and some institutional expedients would appear to offer an economically attractive way for moving through the transitional period.

Much of the nuclear-technology symbiosis strategy that provides the basis for some of the discussions in Chapters IV and V leans heavily on the many papers and speeches by Dr. Peter Fortescue of the General Atomic Company. Acknowledgement is made in the text to the Fortescue literature, but it is impossible to acknowledge adequately the full effect of his outstanding work.

It will also be obvious that the work of Dr. Alvin Weinberg and the Institute of Energy Analysis at Oak Ridge has had a profound impact on directions taken in this book. The IEA work on institutional studies has been particularly stimulating. This, of course, is not intended to suggest that the IEA would endorse or even agree with conclusions of this book.

While this author strongly believes that gas-cooled reactor technologies should be pursued because of some of their unique characteristics, there has also been considerable emphasis in this book on light-water reactors and liquid-metal fast-breeder reactors. That emphasis has been chosen since deployment and development directions have already favored those reactors. To assure a significant impact, then, strategic planning must put major attention on those reactor types, at least for the imminent transitional period.

In summary, it is a hope this book will offer some fresh views on energy problems and policies, with particular emphasis on:

- energy policy time constraints,
- potential energy technology directions, and
- potential energy institutional directions.

As a postscript to this preface, it may be useful to indicate the timeframe of manuscript preparation. Most of the material in this book was prepared in 1979 and the early part of 1980. A few references have been added following completion of the original manuscript, but no substantial changes have been made after early 1980.

In this context, it is interesting to note that very recent total-energy-consumption data for the full year of 1979 showed less than a 1% energy growth over that of 1978; and energy consumption in the first quarter of 1980 appears to be lower than that in the corresponding period of 1979 (data from Department of Energy, Energy Information Agency monthly reviews). While 1980 might be regarded as an abnormal year because of the economic recession, nevertheless, the overall trend in energy consumption appears to be generally consistent with projections of energy growth discussed in the first chapter of this book.

It is anticipated that the continuing decrease in energy growth will lead government and industry to lower their long-range energy consumption projections still more. While this would seem appropriate for the period 1980 to 1995, there is a great danger that a complacency will develop toward technology and institutional planning that could lead to serious problems after 2000. Perhaps the potential danger resulting from such a complacency, might be regarded as one of the important cautionary admonitions of this book.

Part One

ENERGY GROWTH

Chapter I

Energy Growth: Trends, Logistic Curves, Economic Cycles and Crystal Balls

World energy consumption increased almost three-fold from 1950 to 1970. U.S. energy consumption doubled during that period. The production and consumption of oil grew at an even more astounding rate and, at this time, accounts for more than 40% of energy consumption in the world. It is abundantly clear that oil consumption can no longer maintain its historic growth rate. And in the very long range, it is equally clear that essentially all energy production will have to depend on inexhaustible energy resources.

Of fundamental importance is the time interval society has available to make the transition from current energy technologies to the longer-range ones. And, the time interval available for that transition depends critically on the growth rates of both energy consumption and fuel-supply capabilities.

Policy planning for the transitional period must make distinctions between energy-supply problems and oil-supply problems; between U.S. problems and world problems; and between near-term solutions and long-range solutions. Of greatest concern at this time is the oil-supply problem, not energy supply in general. However, a continuing energy-consumption growth and the transfer of energy-consumption patterns from oil to other resources, such as natural gas, coal and uranium, could simply extend the problem to those other resources. Moreover, the larger energy growth rates in other parts of the world and the unavailability of alternative fuel resources in some of those countries will create different kinds of problems for different countries. In all cases, energy conservation appears to be the best alternative for the next few years. But, again, the potential for energy conservation may be quite different for different countries. In the long range, though, inexhaustible energy resources appear to be essential for all parts of the world.

The magnitude of the problems that can develop as the result of mismatches between the demand and supply of energy resources will be subjects to be discussed in Chapters II and III. But, the severity of those problems can only be identified with confidence if the energy growth can be projected with some reasonable reliability. Hence, this first chapter will focus on the very important subject of empirical growth projections--methodologies, comparisons with history and forecasts for the future. Historical data to be used for the illustration of principles will draw heavily from United States statistical information since those data are more readily available.

One interesting conclusion of this examination will be that energy growth has, for more than 100 years, evolved in surges or cycles. If history repeats itself, and reasons will be suggested for such a possibility, energy growth will be slower in the period 1970 to 1995, but another surge will begin in 15 to 20 years. Presuming this conclusion is correct, energy conservation should be achievable in the next one to two decades, and time should be available, at least in some cases, for the development and implementation of technology and institutional redirections to gird for the next surge in energy growth. Potential nuclear technology and institutional redirections to meet that possible surge will be the subject of the last four chapters.

GROWTH TRENDS

During the 20-year period from 1950 to 1970, the growth of domestic energy consumption averaged a robust 3.3% per year. In 1977, the growth had fallen to 2.7%. In 1978, it was 2.1%. Similarly, electricity growth from 1950 to 1970 averaged a remarkable 7.4% per year. In 1977, the growth had diminished to 5.1%, and in 1978 it was 3.7%. It is, of course, risky to attach too much significance to energy statistics over short time intervals, but it would appear that energy consumption is currently showing a decreasing growth trend. Whether this trend marks a success in energy conservation, whether it is a reaction to higher energy prices, or whether it might be attributable to still other more subtle conditions are important questions. Possibly all these factors are contributors, though not necessarily in definable ways.

Energy growth forecasts that have been projected by the government and industry in the last fifteen years have not proven to be remarkable for their accuracy. Just as our typical stock broker is generally optimistic during bull markets and pessimistic during bear markets, national energy planners have shown a proclivity toward this inertial syndrome. In the late 1960s, during the final years of the 20-year high-energy-consumption period, long-range projections for future electrical energy growth were generally bullish. Toward the end of the 1970s when the growth rate of electrical energy consumption had shown a significant reduction, long-range projections for future growth became increasingly bearish, with each new prognosis more conservative than the previous one. One explanation for this apparent change in trends has been that energy growth was encouraged in the late 1950s and 1960s by cheap energy resources, while energy growth in the late 1970s has been discouraged by high resource costs. Undoubtedly that is a factor, but as will be seen subsequently, it may not be the only reason nor even the most important reason for the lower trends in energy consumption.

It has been generally observed by energy policymakers that there tends to be a significant correlation between the gross national product and energy consumption from country to country. How closely coupled that relationship is has been the subject of much argument. Figure 1.1 illustrates the relationship between GNP and energy consumption for a number of countries.[1] While there is considerable scatter in the data, it is apparent that countries enjoying a relatively large GNP also tend to be large energy consumers. However, it should be noted that the highly-industrialized European

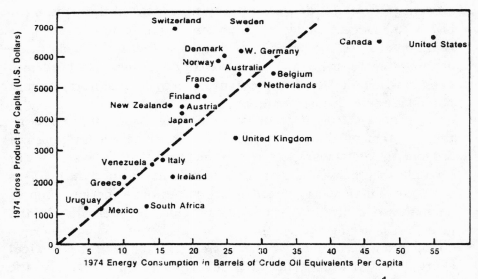

Figure 1.1   Correlation of GNP and energy consumption [1]

countries (particularly West Germany, Switzerland and Scandanavia) have succeeded in achieving a per capita GNP approximately equivalent to that of the U.S. with about half the energy consumption. The strong implication is that some room exists for conservation of energy in this country without threatening our productivity.

Growth curves for our domestic GNP and energy [2,3] are shown in figure 1.2. The growth curves follow each other quite closely, suggesting that rises and falls in energy consumption are reflected by rises and falls in GNP or, perhaps, vice versa. Since employment is related to GNP, it is not difficult to see why industry and labor both view energy consumption as a barometer of economic healthiness.

In spite of the profligate use of energy in the U.S. relative to the rest of the world, there are clear signs that the efficiency of energy utilization relative to GNP is improving. This is illustrated by figure 1.3, where the ratio of GNP/E, i.e., the "productivity effectiveness of energy", is indicated.

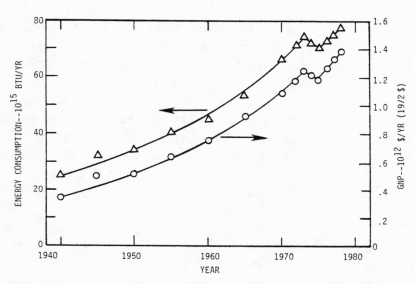

Figure 1.2   Growth curves for domestic energy consumption and GNP

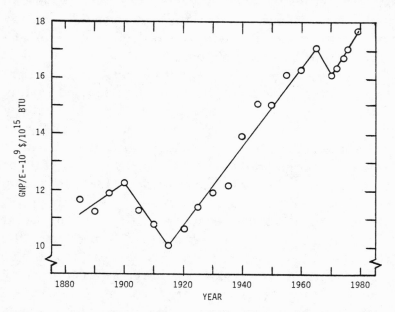

Figure 1.3   Improvement trend for domestic energy productivity

The figure implies that the energy productivity improved steadily from 1915 to 1965 (even through the great depression), decreased briefly from 1965 to 1970, and resumed its increase after 1970. It is not clear why the energy productivity fell in the five-year period from 1965 to 1970. Perhaps it can be attributed to the fact that the cost of energy in constant dollars reached a minimum during that period--although the difference in energy cost over that for the preceeding ten years was very small. Moreover, the decrease was too early to be blamed on environmental constraints which began around 1970.

It is probably more interesting that the improvement rate for energy productivity shows approximately the same slope following 1970 as that from 1915 to 1965. This would suggest that the higher costs of energy and the emphasis on energy conservation in the last few years have had a beneficial effect that just about offsets the incremental energy commitments arising from environmental control and the recovery of lower-grade energy resources.

While rising energy costs are still not as great as might be expected in future years--a subject to be discussed further in the next chapter--the apparent trends in energy productivity improvement are not sufficient to lend great enthusiasm for the accomplishments of conservation to date. If improvements in energy productivity continue at the rate of 1% per year, as was the case from 1915 to 1965, it would require another 70 years to double our domestic efficiency, i.e., reach a level approximately equivalent to that of West Germany and Sweden. With improvements of 2% per year, the doubling time would be 35 years, and at 3% per year, it would be 25 years. The National Research Council report of the committee on nuclear and alternative energy systems (CONAES) [4] concluded that technical efficiency measures alone could improve the energy productivity by as much as a factor two over the next 30 to 40 years. Clearly, then, a much more dedicated effort toward energy conservation is required, if we are to make significant progress in that direction. The more fundamental questions are:

- at what reduction rates in energy consumption might one expect deleterious effects on the economy; and
- what level of growth change might be reasonably expected for our society.

Those questions and related ones can be put in somewhat better perspective by examining historical and projected growth trends for energy consumption.

LOGISTIC CURVES

Energy growth can be projected by any of at least three general approaches, viz., those involving:

- econometrics,
- end-use integration, or
- trend analysis.

The first approach is basically analytical, the second integral, and the third empirical. Some further distinctions between the alternative approaches are discussed in Appendix A, with particular attention given to logistic methodology. Reference 5 (at the end of this chapter) is recommended as a good example of a study using econometric growth projections, while references 6 and 7 are particularly interesting as examples where end-use-integration projections have been used. Each of these approaches is quite tedious, but they have the merit of relating energy growth to anticipated changes in population, work productivity, gross national product and energy costs. Both the end-use-integration and the econometric methods are generally expected to yield reasonably reliable results for forecasts involving 10, 20 or, perhaps, 30 years forward. As with all projection methodologies, the reliability decreases as the time span of the forecast increases.

An alternative, more simplistic approach is one using an empirical methodology. The empirical approach, to be used in this and succeeding chapters, leans basically on the application of logistic growth curves, chosen to fit historical data appropriately. Possible patterns of deviation from the logistic growth are then sought with the final objective of predicting future patterns that can be superposed as perturbations on the expected continuing logistic growth. Perturbations arising from economic cycles, for example, will be examined in some detail in the next section.

The justification or, perhaps, rationalization for using an empirical projection methodology in this analysis is based on:

- its relative simplicity,
- the very long projection periods to be examined, and
- the observation that energy and economic growth curves tend to follow very stubbornly certain natural growth laws.

Obviously, the application of this simple methodology is subject to considerable uncertainty, particularly in the event of major social disorders, such as economic reversals, international instabilities and major wars, to mention only a few. But, at least some of those disorders could produce similarly disruptive effects using the other methods of projection. Some of the uncertainty in the empirical approach might also, perhaps, be minimized by requiring it to conform somewhat reasonably with the results of the more sophisticated approaches if, indeed, significant differences should appear.

The logistic or sigmoid growth curve has commonly been used in the study of biological growth, demography and, occasionally, economics or business systems. The logistic growth function is characterized by well-defined constraints on the initial and terminal boundary conditions. For example, the population growth of fruit flies in a closed food-supply system begins by following an exponential growth, but as population reaches a point where available food resources become limiting, the growth rate decreases exponentially, with the population now approaching an asymptotic upper limit. Mathematically, the logistic growth is based on the fact that fractional growth rates are proportional to an exponentially-saturating parameter.* The logistic or sigmoid curve typically has a shape somewhat as follows:

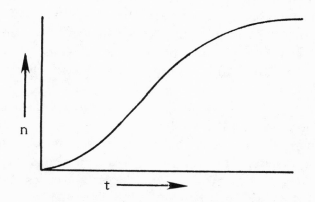

Necessarily, a logistic growth must assume that no abrupt changes in the terminal boundary conditions can occur. For example, it assumes that the cost of food, energy or other essential supply resources would increase regularly as the resources diminish. In the case of energy resource supply it would presume that alternative resource systems could become available, at not-greatly different costs, in the event one resource approaches depletion.

While the *overall* energy supply rate might be expected to follow a logistic growth, the supply rate of a single depletable fuel resource, such as coal or oil, would grow rapidly in the early years of utilization; would approach a

*See Appendix A for more discussion on this.

maximum utilization rate during the zenith of its popularity; and finally would decline to zero as that resource became depleted, or its price became economically unattractive. It is possible, as might be the case for coal, that a particular fuel could be displaced by a cheaper alternative temporarily, but could enjoy a resurgence in utilization when the alternative becomes uncompetitive or insufficient to meet all demands.

It is even conceivable that the logistic curve for energy growth as a whole might be altered by a major change in resource availability or a breakthrough in technology. Hence, one must be cautious in attaching too much fervor to the doctrine of logistic growth. Nevertheless, the concept of logistic growth will be accepted here as one plausible growth pattern for both total energy growth and electricity growth.

Data exist for total domestic energy consumption covering a period from 1850 to 1978 and electricity energy consumption from 1905 to 1978.[2,3] Since those periods cover several wars, a number of business cycles and shifts in fuel resources, one might expect that the logistic curves fitting those historical data would, at least, supply a rough indication of growth for, another 50 years. Figure 1.4 illustrates selected logistic growth curves for total energy, electricity energy input and electricity supply, all expressed in kwh (thermal), or kwh (electric) per annum. The actual comparison of historical data and the logistic fits for total energy and electricity energy will be discussed in more detail in the next section.

Obviously, there is some hazard in selecting a unique fit to data over a period of some 100 years. In addition to the problems associated with social disorders and the possibility of changing terminal constraints previously indicated, the data reflect the combined effect of population growth curves and per-capita, energy-consumption growth curves.

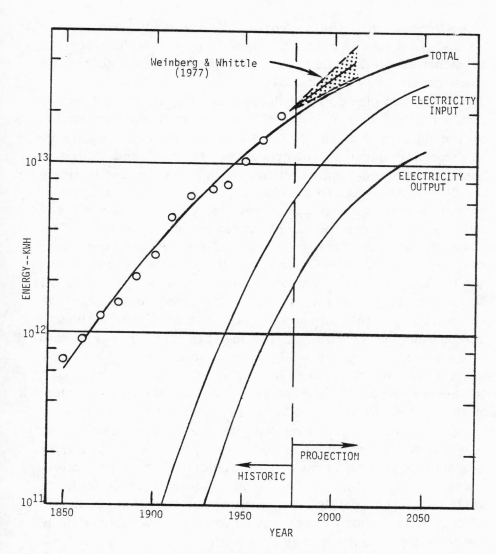

Figure 1.4   Logistic curves for domestic energy growth

Nevertheless, with the exception of some cyclic behavior to be discussed subsequently, the data do appear to be amenable to rather simple logistic growth trends over the complete time spans covered by historical data.

The top curve in the figure indicates the total domestic consumption of energy from 1850 to 1978 with an extrapolation of the data to the year 2050. The triangular band just above the curve illustrates projections by Weinberg and Whittle [6] based on the end-use-integration studies done at the Institute for Energy Analysis. The median curve follows the forecast of total domestic energy projected in 1978 by the Energy Information Agency of the Department of Energy. [8]   The slope of the median growth appears to be very similar to that of the logistic curve.   However, the logistic growth curve would suggest that actual energy consumption in 1975 was about 15% above that of the indicated logistic growth projection.   If this is correct, and if that abnormal energy consumption must return to the logistic, then energy growth for the next few years would have to follow more nearly the lower-limit growth projection of Weinberg and Whittle.   But, further discussion of that apparent anomaly will be the subject of the next section.

It is also interesting to note that the logistic curve extrapolates to about 9×10^{13} kwh/yr or 10^{10} kwt-yr/yr around the year 2100 A.D. This would suggest a per capita energy consumption of 20 to 25 kwt-yr/yr, the resulting value depending partly on the assumption used for population at that time.   That extrapolation compares favorably with an asymptotic per-capita energy consumption of 20 kwt-yr/yr projected by Weinberg [9] and occasionally used by others.

The energy input curve for electricity generation has been calculated assuming a continuing rise in generation efficiency following somewhat historical trends. For example, the overall system efficiency was 17% in 1930, 24.3% in 1950 and 32.5% in 1970. It is projected at 35% in the year 2000 and around 40% in the year 2050.

On the basis of historical data, the energy input used for electricity generation was approximately 10% of the total domestic energy consumption in 1930, was almost 20% in 1950 and about 25% in 1970. By 1980, the fraction should exceed 30% and the logistic curves indicate fractions of about 45% in the year 2000 and around 70% in 2050. These projections are somewhat more optimistic for electricity generation than some other projections and, indeed, such long-range projections can err significantly. But, the present trends in electricity and total energy consumption rates do indicate a continuing preference for the convenience of electricity as a source of energy.

ECONOMIC CYCLES

It has already been suggested that energy might grow in surges instead of monotonically. Should that be the case, it might be useful to explore more precise patterns of deviation from the normal growth curve.

Figure 1.5 illustrates the historical growth of total energy consumption in the U.S. over the period 1850 to 1975, with the logistic curve shown. The data points, shown by circles in the figure, are five-year moving averages at five-year intervals. Data were used from Department of Commerce and Department of Energy [2,3] sources. An examination of the figure shows that the data points tend to have a cyclic variation about the logistic curve.

Figure 1.6 illustrates similar data for the growth of electricity consumption from 1910 to 1975. Again the data points tend to fall above the logistic curve prior to 1930, below the curve from 1930 to 1950, and above the curve more recently. Since the energy consumption is shown on a logarithmic scale in both cases, the deviation from the normal growth curve may appear deceptively small.

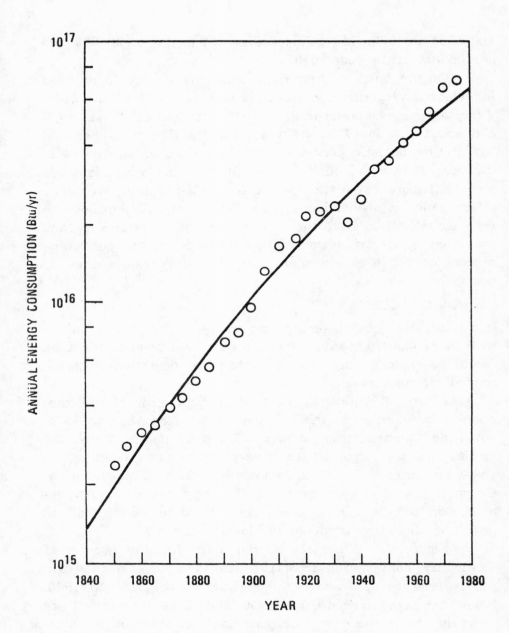

Figure 1.5   Growth of U.S. total energy consumption

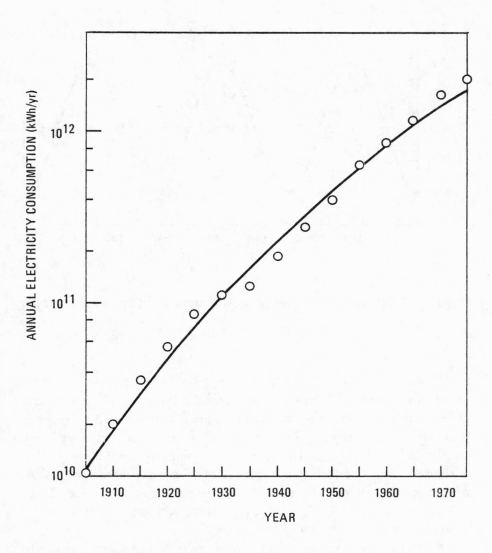

Figure 1.6   Growth of U.S. electricity consumption

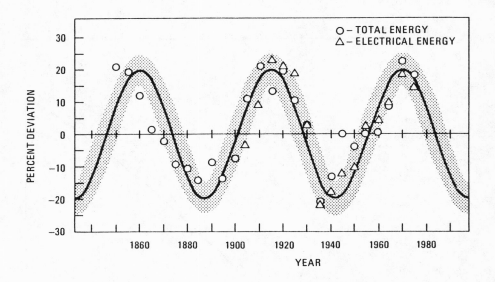

Figure 1.7 Deviation of energy growth from a logistic approximation

        In figure 1.7, the percentage deviations of individual
points from the logistic curve are shown for both total energy
and electricity energy consumption. A cosine curve with an
amplitude of 20% and a period of 55 years is also shown with
a shaded area bordering the curve. The shaded area covers an
amplitude band from 15 to 25% at the peaks and a time band
± 5 years in period. Quite remarkably, almost all the energy
deviation points fall within the band.
        Assuming such a variation in energy consumption could
persist, it is interesting to examine how this variation might
affect both past and future energy projections. Figure 1.8
illustrates the logistic growth of electricity from 1930 to
2040, and a "cycle-adjusted-logistic" (CAL) curve obtained by

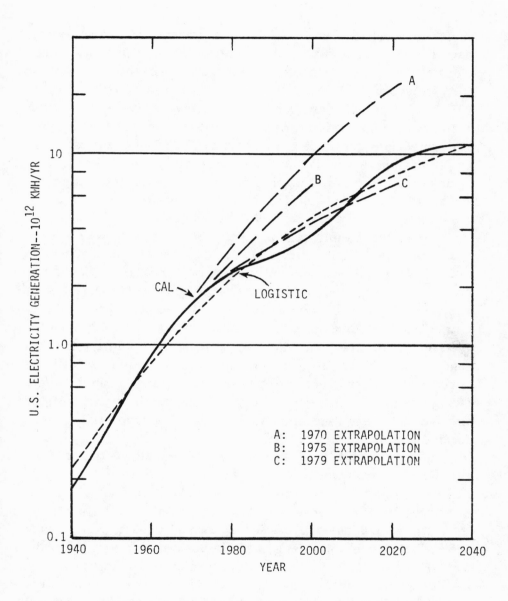

Figure 1.8   Projected growth of U.S. electricity consumption

the superposition on the logistic curve of a sinusoidal deviation having a 20% amplitude and a 55-year period. Three plausible extrapolations from the curve are shown assuming the projector is looking back from the year 1970, 1975 or 1979. Those extrapolations are roughly equivalent to AEC projections[10] in 1971, an EEI projection[11] in 1975 and typical projections now being forecast by DOE[8] and others. Just as projections A and B have subsequently proven to be overly optimistic, there is a danger now that projections for the next 10 to 15 years could seriously underestimate a growth surge that could appear between 1995 and 2025.

The remarkably regular cycles in energy growth raises specters once again of the unpopular concept of a long economic cycle. In particular, since the character of GNP behavior tends to follow that of energy, it might be expected that economic behavior would follow the same cyclic trend. Considerable data exist on mineral recovery rates, on commodity production rates, on wholesale prices and on interest rates suggesting, indeed, the presence of a long wave in business activity.[12] However, the concept of a long business cycle has generally been rejected by economists, both because of the lack of clear data and the absence of a supporting theory for the cycle.

The possibility of a long wave is usually attributed to the Russian economist and professor, Kondratieff.[13] The general rejection of the Kondratieff cycle is reflected somewhat by the adversities that personally befell its author. In articles published from 1922 through 1925, Kondratieff showed evidence for a long wave in business activity over the period from 1700 to 1920 and predicted a depression in capitalist countries beginning in the late 1920s. One might have thought the projection of a depression in capitalist countries would have been greeted with some enthusiasm by the Marxist government. The implication, however, that a business depression in capitalist countries would ultimately be self-correcting proved to be the downfall of Kondratieff in Russia. Subsequently, he was banished to a concentration camp in Siberia where he died in obscurity years later.

When the great depression, beginning in 1929, seemed to confirm Kondratieff's predictions, the long-wave theory caught some temporary attention in the western world. However, with the advent of Keynesian economics and other indications that business cycles could be at least partially controlled by economic policies, Kondratieff fell into disfavor there also. Indeed, one might speculate that Kondratieff lost favor in Russia because he suggested a recovery after a depression, and he lost favor in the western world because he suggested a depression after a recovery.

The Kondratieff long-wave theory was pursued by the Harvard economist Schumpeter in his book on business cycles.[14] Although new technology implementation was already suggested by Kondratieff as a basis for the initiation of long waves, Schumpeter pursued this argument much more extensively and systematically. The basic argument was that the upswing of a new long cycle generally resulted from the commercial introduction of new technologies. Following a period of economic depression, the newer technologies presumably injected economic momentum into a system that had faltered as the result of previous technologies becoming exhausted. It was the theory, then, that the new technology introduction provided a stimulus for a renewed surge in economic growth. But, after some substantial growth, including possibly some overgrowth, markets became saturated and a decline in business activity marked the end of the growth period. Just as the initial growth momentum was viewed as a regenerative phenomenon, so also was the declension. The resulting stagnation of growth led inexorably to another economic depression.

While the growth periods could, in fact, be correlated with the predominant growth of some new technology, the cause and effect relationship was difficult to ascertain and the theory was at best a hypothesis. Except for rather vague statistical data, then, the long cycle has generally lacked credibility, at least until recent years. Within the last few years, the subject has again been attacked by Forrester [15] of the MIT Sloan School of Business Management and Rostow [16] of the University of Texas. Kahn [17] in his recent book on international economics also reviews some of the new activities on the long wave.

Probably the most interesting and relevant work is that of Forrester. In several articles dating from 1976, Forrester has described work being done on economic cycles using the Systems Dynamics National Model developed at the Sloan School. The model is a laboratory representation of the U.S. economy that simulates the flow of people, information, goods, prices and money. It contains some 15 separate industrial sectors, such as consumer durables, capital equipment, energy, agriculture, etc. Each sector actually simulates a typical business in that sector. The model, then, performs the various steps in business activity, handles financial operations, represents the supply and demand of the marketplace, allows for labor mobility and accommodates government policies. With the model, the economic behavior of the entire economy can be examined.

Perhaps one of the most interesting results of the model studies has been the observation of cyclic variations corresponding to the 15- to 25-year Kuznets cycle and the 45- to 60-year Kondratieff cycle. Although Forrester notes in his papers that the model was not developed for the purpose of studying long-wave economic behavior, initial studies on the interaction of the consumer-durables sector and the capital-equipment sector showed a very strong cyclic behavior of about 50 years. Indeed, it was their observation that the long-wave was a more dominant cycle than the shorter Kuznets cycle. A study of the model has shown that the long-wave results from an intensive expansion of the capital sector. This, in turn, forces major labor transfers to that sector creating a generally healthy expansion of the economy. As the capital expansion grows, the addition of new facilities ultimately exceeds the needs of the industrial market. When the expansion declines, then, the labor market in this segment diffuses to other industry segments creating an overall surplus of labor. Ultimately, the system collapses of its own weight and the economy must then await either the obsolescence of the existing facilities or a new capital growth surge resulting from the introduction of new technologies with more promising economics. Since the average life of capital equipment tends to be around 40 to 50 years, the cycle is enhanced by this replacement necessity.

Another interesting aspect of the model studies is the opportunity it affords for examining the effectiveness or leverage of various policies in applying corrective action to the economy. While the reader should refer to the references [16] for details on this subject, it is worth noting that the low-leverage policies studied by the model were normally policies that attacked the symptoms of a problem, while the high-leverage policies attacked root causes of the problem.

If, indeed, the long wave results largely from the obsolescence of capital equipment and, perhaps, technologies themselves, it would seem obvious that policies should be encouraged at the appropriate times to expedite the introduction of new and more economically attractive technologies. In view of the serious problems arising from the squeeze on oil resources, energy technology would seem to be an inviting area for the application of high-leverage economic policies.

CRYSTAL BALLS

While the subject of cycles is an interesting and provocative one, the acceptance of the inevitability of cyclic behavior--or even an uninterrupted logistic energy growth--is not essential to the basic argument to be presented in this book. Nevertheless, it is already clear that both total energy consumption and electricity energy consumption during the last ten years are growing at a decreased rate relative to the preceding 30 years. Furthermore, on the basis of forward orders for generating capacity, it seems likely that electricity (and probably total energy growth) will continue to grow at a more modest rate for at least another 10 to 15 years. Probably, then, the assumption of a cycle-adjusted-logistic growth for the next 15 years, at least, is not unreasonable, whether this change comes from higher costs of energy, from energy conservation efforts or from a cyclic phenomenon itself.

Figure 1.9 illustrates the cycle-adjusted-logistic growth curve for both total energy and electricity for the period from 1920 to 2040. Assuming that the amplitude of cyclic variation is 20%, then the curve for total energy consumption between 1980 and the year 2000 would actually suggest a zero growth (indeed, between 1980 and 1990 the total energy consumption would decrease very slightly). Because of the higher logistic growth for electricity, the cyclic effect would not override this growth. However, the growth rate of electricity between 1980 and 2000 would be about 2.5% per year; considerably below the 7.5% per year for electricity growth between 1950 and 1970. Energy growth, therefore, would not appear to be the fundamental U.S. energy problem for the next 20 years. More importantly, oil consumption and our oil dependence will continue to be the basic problem.

If energy growth does, indeed, continue to follow the assumed cycle-adjusted-logistic curve for a period of some 50

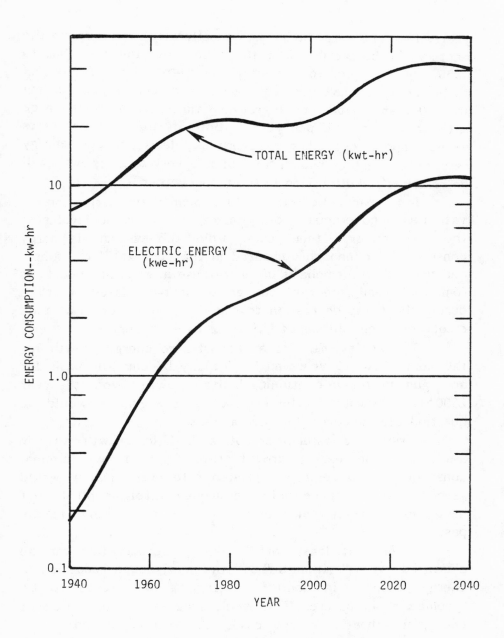

Figure 1.9   Cycle-adjusted-logistic growth curves for U.S. total
and electricity energy

years, then the surge in growth shortly after the turn of the century will be enormous. In the period from the year 2000 to 2030, for example, total energy would grow by a factor 2.3 times, or about 2.8% per year. Electricity consumption during this 30-year period would grow 3.6 times, or at an average rate of almost 4.5% per year. Both of these growth rates would impose exceedingly severe problems on our energy resources. Further implications of this very large growth will be reserved for discussion in the next chapter.

In summary, then, it would appear that energy conservation as a government policy might be readily achievable in the next ten or fifteen years. While the wisdom of energy conservation cannot be disputed in the context of oil supply and demand, the emphasis on energy use in general, though, is somewhat reminiscent of the government-mandated programs during the great depression to reduce the production of food resources when millions of U.S. citizens were hungry.

The achievement of an almost-zero energy growth for the next 10 to 15 years might actually become the basis for an argument that a continuing limited growth beyond the year 2000 is achievable. The logic of that argument would be questionable, however, for several reasons.

First, the assumption of a logistic growth already assumes a decreasing growth rate imposed by terminal constraints. To require a growth rate even smaller would seem to require regimented government controls--controls that have neither been remarkably successful nor popular in the past.

Second, at least part of energy conservation can be classified as an "arithmetic" improvement, i.e., a reduction in energy losses. In contrast, energy growth tends to be "geometric" in nature. This would suggest a continuation of energy growth after energy losses have been minimized.

Third, increasing energy requirements to assure environmental cleanliness and to develop mineral, water and food resources under more difficult future conditions will probably offset some of the improvements that might otherwise be accomplished in energy productivity.

Fourth, a new growth in industry, following the obsolescence of our present capital equipment, could place heavier demands on energy.

Perhaps most important of all, it is likely that the growth of living standards in other parts of the world will have some impact also on the already-industrialized nations. If the U.S. and other industrialized nations are to provide some leadership in improving the living standards throughout the world, it will almost certainly require some extra energy expenditure--possibly for reasons other than improving our own gross national product.

For all the foregoing reasons, energy growth in the next century will probably follow at least the rate suggested by a logistic curve, and possibly more nearly that of the cycle-adjusted-logistic curve. The more important questions will probably center on types of energy resources rather than on energy consumption per se--at least after some period of apparently successful energy conservation. For purposes of technology planning, then, it would seem prudent to assume an energy growth in the 1995-2025 period following that of the cycle-adjusted-logistic curve, again recognizing that energy costs, energy conservation and economic planning could, however, mitigate some part of the large energy growth that would normally be projected by the CAL methodology.

REFERENCES

1. The National Energy Plan, Executive Office of the President, April 29, 1977.

2. "Historical Statistics of the United States-Colonial Times to 1970", U.S. Department of Commerce, Bureau of the Census, Bicentennial Edition, Parts I and II, September 1975.

3. "Monthly Energy Review", U.S. Department of Energy, Office of Energy Data, June 1979.

4. "Energy in Transition, 1985-2010", Final Report, National Academy of Sciences, 1979.

5. Nuclear Power Issues and Choices", Report of the Nuclear Energy Policy Study Group, Ballinger Publishing Company, Cambridge, Mass., 1977.

6. Weinberg, Alvin M., and Whittle, Charles, E., "Energy Policy and Projections: The Case of a Nuclear Moratorium", Nuclear News, March 1977, pp. 44-48.

    Weinberg, Alvin M., et al, "Economic and Environmental Impacts of a U.S. Nuclear Moratorium, 1985-2010", Institute for Energy Analysis, The MIT Press, 1979.

7. "Energy Supply-Demand Integrations to the Year 2000", Third Technical Report of the Workshop on Alternative Energy Studies, The MIT Press, 1977.

8. Annual Report to Congress 1978, Energy Information Agency, DOE/EIA-0172, 1978.

9. Weinberg, Alvin H., Hammond, R. Phillip, "Limits to the Use of Energy", American Scientist, July-August 1970, Volume 58, Number 4, pp. 412-418.

10. "Cost-Benefit Analysis of the U.S. Breeder Reactor Program", U.S. Atomic Energy Commission, Division of Reactor Development and Technology, updated (1970), January 1972, WASH-1184.

11.   "Economic Growth in the Future", Edison Electric Institute, Mc-Graw-Hill Book Company, 1976.

12.   Dewey, Edward R. and Dakin, Edwin F., "Cycles-The Science of Prediction", Henry Holt and Company, New York, 1947.

13.   Kondratieff, N.D., "The Long Waves in Economic Life", The Review of Economic Statistics, Volume XVII, Number 6, November 1935.

14.   Schumpeter, J.A., "Business Cycles", McGraw-Hill Book Company, New York, 1939.

15.   Forrester, Jay W., "Growth Cycles", De Economist 125, NR. 4, 1977.

      ------"Business Structure, Economic Cycles, and National Policy, presented at the National Association of Business Economists, Florida, 1975, Futures, June 1976.

      ------"A Great Depression Ahead?-Changing Economic Patterns", The Futurist Magazine, December 1978.

      ------"A Self-Regulating Energy Policy", Keynote Address of the AIAA 15th Annual Meeting, Astronautics & Aeronautics, July-August 1979.

16.   Rostow, W. W., "Getting from Here to There", McGraw-Hill Book Company, 1978.

17.   Kahn, Herman, "World Economic Development-1979 and Beyond", Morrow Quill Paperbacks, New York, 1979.

Chapter II

Fossil Fuels: The Perils of Paucities

In his farewell address to the Department of Energy, August 1979, Schlesinger said "Quite bluntly, unless we achieve the greater use of coal and nuclear power over the next decade, this society may just not make it." [1]   As a former cabinet officer of both the Department of Energy and the Defense Department, Schlesinger had been in an unusually good position to assess the energy problems of this country and their implications on world stability. Schlesinger further emphasized "The worldwide system for the production and distribution of petroleum is already stretched taut. There is little, if any, relief in prospect. Any major interruption--stemming from political discussion, political instability, terrorist acts, or major technical problems--would entail severe disruptions."   That grim warning preceded the Iranian seizure of the U.S. Embassy and the Russian invasion of Afghanistan late in 1979.

In a 1977 study, the Central Intelligence Agency predicted that the USSR would become a net importer of oil in the mid-1980s. [2]   While the report was very controversial at the time, subsequent events seem to confirm that probability. Already, the Soviet supply of oil to the Comecon countries has been limited, forcing them also to seek additional supplies from OPEC. [3]   With this as a backdrop, it is not surprising to see the Communist countries also casting an anxious eye toward the Middle East oil supplies. If Soviet influence should be extended to the Middle East, then Schlesinger points out that "would mean the end of the world as we have known it since 1945 and of the association of free nations".

Adding also to the problems of the U.S. and the other industrialized free nations are the impending energy problems of the evolving industrial nations. In some sense, the energy squeeze poses a more severe threat to the growth of the developing countries than to that of the wealthier nations."

A basic objective of this chapter is to examine probable energy growth patterns in the U.S. and the world, putting particular attention on resource supply/demand imbalances that might occur if attempts should be made to rely primarily on the use of diminishing supplies of fossil fuels. Two recent studies are especially relevant to this subject.

One study conducted by the MIT Workshop on Alternative Energy Strategies (WAES) [4] focuses on the world energy problems with particular attention on the oil supply/demand problem. It observes that world demand for oil will exceed the growth of oil-producing capacity in the 1980s without some relief from alternative energy resources. It concludes, then, that the continuing economic growth of the world will depend on a major transfer of emphasis from the current predominant use of petroleum and natural gas resources to the much greater use of nuclear energy and coal by the year 2000.

The other study by the Energy Project of the Harvard Business School [5] limits attention primarily to the domestic energy problems. It concludes that a key energy resource for this country should be energy conservation and that supplementary energy requirements can be largely satisfied by solar technology. As will be seen subsequently, the Harvard Energy Project conclusion may be a valid one for the U.S., at least up to the year 2000; but the WAES conclusion is probably more realistic for the world in general. Moreover, it appears that the WAES emphasis on nuclear and coal resources will become increasingly important for both the world and the U.S. beyond the year 2000.

Currently, the "energy crisis" in the U.S. is basically an "oil crisis". However, for most of the industrialized nations, the oil crisis is compounded by the lack of other indigenous energy resources. In 1978, the consumption of oil in the U.S. accounted for almost 50% of this country's total energy consumption. Of this, almost 50% of the oil consumption was met by imported oil. In effect, some 25% of the U.S. energy supply depended on the import of energy resources.

While the energy dependence of the U.S. is disquieting, the energy dependence of Japan and some of the European countries is even more unsettling. In France, for example, more than 75% of their energy resources must be imported. [6] With a very strong program to utilize conservation, solar energy and, particularly nuclear energy, the French hope to reduce their dependence on fuel imports to less than 60% in the next 20 years. For western Europe in total, about 65% of the energy resources must be imported; and for Japan the fraction is around 90%. [7] Japan and Germany also have plans to reduce their energy dependence, primarily through the greater utilization of nuclear energy. England is somewhat better off because of the oil discoveries in the North Sea, but even in that country there is an effort to minimize the dependence on petroleum. Obviously, the energy crisis is not a domestic crisis. It is a problem of international dimensions with the stability of the world depending on a solution that allows greater energy security and independence.

Since 45% of the world's energy requirements and about 50% of the U.S. energy requirements are currently met by petroleum-based fuels, the energy crisis clearly involves dependence on oil at this time, both in the U.S. and the rest of the world. However, it is likely that the crisis will soon extend to coal as the emphasis is shifted to its use, thereby putting a strain on the world supply of this fossil fuel, perhaps within 20 to 30 years. Indeed, the energy crisis will probably become a crisis of supply for all fossil fuels in the not-too-distant future. While 50% of the U.S. energy consumption depends on oil, for example, some 75% depends on a combination of oil and natural gas, both of which are expected to diminish in availability over the next few decades. In fact, including coal, 92% of our domestic energy consumption depends on fossil fuels. But, at least the U.S. derives some 75% of its fuel resources from within its own territory--in sharp contrast to a figure of around 25 to 35% for most European countries and about 10% for Japan.

Since over 90% of the world's energy supplies also come from fossil-fuel resources, the implications of further energy growth on the availability and use of these resources must be examined. In this chapter, then, the following subjects will be pursued:

- the growth of U.S. and world energy demands;
- the energy-demand implications on resource availability;
- the energy-demand implications on energy independence; and
- the energy-demand implications on environmental impact.

Primary emphasis throughout Chapter II will be on the role of fossil fuels in the supply/demand problems. The role of nuclear fuels will be discussed in Chapter III.

GROWTH OF U.S. AND WORLD ENERGY DEMANDS

Energy growth patterns in the U.S. have already been discussed briefly under the subject of logistic growths. Attention here will be focused more on the broader problems associated with world energy growth. An important conclusion of this examination will be that national energy-policy planning cannot ignore the critically important energy-growth patterns that will evolve in the rest of the world.

The importance of world energy consumption can be seen from figure 2.1 which shows logistic and cycle-adjusted logistic energy-demand projections for both the U.S. and the world. It is generally recognized that the U.S. is a profligate

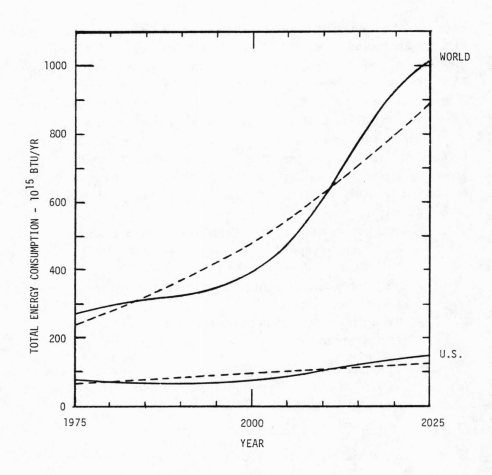

Figure 2.1   Logistic and cycle–adjusted–logistic energy
            consumption projections

consumer of energy. With about 5% of the world population, the U.S. accounted for some 30% of the world energy consumption in 1975, and almost 40% of the energy consumption in the non-communist world. Assuming the projections shown in the figure, the U.S. will account for only 15% of the world energy consumption by the year 2025. While energy consumption in the U.S. can be expected to double in this 50-year period, the world energy consumption will quadruple.

Since the generalized logistic and cycle-adjusted-logistic energy projections will be recalled frequently throughout the book for energy-demand examinations, it will be useful to compare results from these projections with the results of more traditional projections. Comparisons for the projections of U.S. energy growth are indicated in table 2.1. As can be seen, the cycle-adjusted-logistic projection leads to an energy consumption in the year 2000 that is conspicuously lower than traditional projections. In contrast, the CAL projection for 2025 is considerably closer to traditional values. One possible exception, is the CONAES (Committee On Nuclear and Alternative Systems)[8] projection that brackets the CAL projection over the 35-year period covered by that study. And, in fact, an extrapolation of the two median CONAES cases would suggest a U.S. energy consumption of about 155 quads in 2025, a value fairly close to the 147 quads indicated by the CAL projection.

Admittedly, the CAL projection of 74 quads for the year 2000 appears to be unrealistically low. This results from the combination of an assumed cyclic downtrend and an energy growth that already is approaching saturation in the U.S. Although the CAL projection for growth attentuation goes far beyond most other projections, such a prospect is not inconceivable. And by projecting at least some decrease in energy-consumption before it occurs, perhaps the projection of

Table 2.1   Comparisons of projections for U.S. energy consumption

	1975	2000	2025
ERDA–IERS (1977) [9]	74	118	155
DOE–EIA (1978) [10]	75	125	174
WAES (low case) [4]	75	115	
CONAES (1979) [8]	75	63–135	
Logistic	65	94	125
Cycle–Adjusted–Logistic	74	74	147

a subsequent energy resurgence will be more plausible after energy consumption has apparently reached a zero or negative growth.

Table 2.2 shows similar data for world energy projections, though the comparison is more difficult because synthesized data are not so readily available. Some rounded numbers are shown for "typical" projections that tend to be consistent with a number of sources. [4, 11, 12, 13]   Here the logistic data are in good agreement with the typical projections and even the cycle-adjusted-logistic projections may not be so controversial, in spite of the 20% lower CAL projection for 2000.

The fact that the CAL projection is not so startling for world energy consumption in 2000 simply reflects the fact that a cyclic variation on a high-growth part of the logistic curve does not appear to be so dominating. As previously emphasized, the cycle-adjusted-logistic assumption will be given some attention in the following discussion largely

Table 2.2   Comparison of projections for world energy consumption

	1975	2000	2025
Typical	250	500	950
Logistic	237	486	891
Cycle–Adjusted–Logistic	277	393	1069

because it happens to reflect contemporary conditions (though it may overstate the growth slowdown for the next 10 to 20 years). It should be recognized, nevertheless, that the energy problems would appear to be more severe, between 1980 and 2000, should energy growth approximate more nearly a uniform logistic growth.

The implications of the energy consumption data in tables 2.1 and 2.2 can be visualized somewhat better from figure 2.2. In this figure, the cycle-adjusted-logistic energy consumptions for the U.S. and the world are shown for 1975, 2000 and 2025 by bar graphs. For the period 1975 to 2000, a zero energy growth is indicated for the U.S., but a 40% growth is implied for the world. This actually suggests a 60% growth for the world outside the U.S., or almost 2% growth per year. In fact, the 25-year change in energy growth for the rest of the world will apparently be greater than the total energy consumption in the U.S. in 2000.

Again, the fact that the model shows no change in energy consumption for the U.S. between 1975 and 2000 should not be the basis for a condemnation of the general argument.

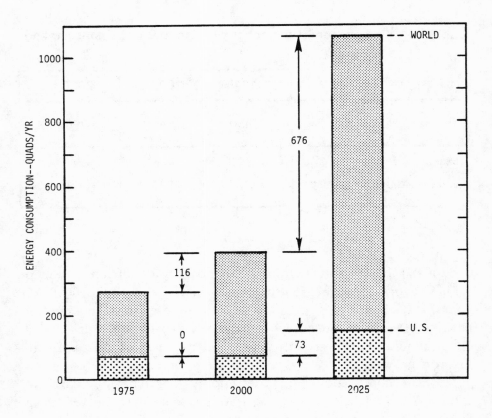

Figure 2.2   CAL energy consumption projections in the U.S. and
               the world

The more important point is that energy growth in the rest of
the world will undoubtedly continue at a relatively brisk rate
even if energy growth is essentially zero during this period
for the U.S.   Energy resources must be found to support the
world growth although energy conservation might be a
spectacular (though possibly an unwelcome) success in the U.S.
This simply emphasizes the point that the energy crisis is
more accurately a world crisis--not a U.S. crisis.

Examining the projected energy growth from 2000 to 2025, it would seem that the energy crisis will reach enormous proportions at that time. While the U.S. energy consumption will increase by 73 quads during that 25-year period, a growth of 2.8% per year, the world energy consumption will apparently increase by a staggering 676 quads--an increment almost an order of magnitude greater than that of the U.S. While a world energy growth of 4% per year would, in itself, appear to be an astounding growth rate, the more important question is one of finding some 600 quads of incremental energy supply when the world is already struggling to increase the oil supplies of about 100 quads by a few percent in 1979-1980.

Three courses of action are open to the world. One alternative is simply to ignore the problem with the tacit intent of establishing an energy status quo, allowing the U.S. to enjoy their present level of energy utilization, but depriving the rest of the world that opportunity. The second alternative, which is an intermediate course of action, is to make some of the U.S. energy resources available to other countries through an austere domestic energy budget, including a much more efficient utilization of energy (i.e., energy conservation) and possibly some successful deployment of solar energy. It would appear, though, that this course of action would have very little impact on the world energy problem, particularly in the second 25-year period. This again would probably require the rest of the world to be content with a per-capita energy consumption considerably below that of the U.S. The third alternative is to use the 25-year period from 1975 to 2000 to develop technologies that can be satisfactorily deployed throughout the world in the period 2000 to 2025 to satisfy the enormous growth that might be anticipated at that time.

Based on energy technologies currently available, it would seem that nuclear energy is the only assured candidate for satisfying the very large incremental energy requirements indicated for the period 2000 to 2025. While a decision for the further *deployment* of nuclear energy may not be necessary at this time in the U.S., it is clear that whatever energy technology might be adopted after the turn of the century must be ready for major implementation around the year 2000. However, the need for that decision may be considerably earlier for the rest of the world. It would not be prudent, therefore, for the U.S. to discourage that decision in other countries, in spite of any needs or lack of needs for nuclear expansion in this country.

Energy Demand Implications on Resource Availability

During the last 25 years, most of the energy growth in the U.S., and also the rest of the industrialized world, has leaned heavily on the increasing utilization of oil and natural gas fuel resources. Not only will prospects for further growth of these fossil fuels be minimal in the next few decades, but it is likely that supplies from these resources will actually diminish. Hence, in the U.S., transfers to the use of alternative energy resources should begin during this period. Both because of the greater energy dependence and the greater energy growth in the rest of the world, transfers to other energy resources will probably be even more crucial there.

It has been emphasized that the energy crisis at this time is primarily a problem resulting from a petroleum supply/demand imbalance; but, the resolution of that problem may create other resource supply/demand imbalances. The liklihood of imbalances for yet other resources can be put in perspective by examining the energy-supply deficit that must be filled by coal and other resources, assuming no further world growth in petroleum and natural gas supplies. Such an exercise is, admittedly, only qualitative, but the results are, nevertheless, illustrative. A cursory analysis of that kind, then, will be pursued here; both for the U.S. and the world, and both for the 1980-2000 period and the 2000-2025 period.

In the U.S., approximately 55% of the oil consumption is currently used for transportation and another 10% for petro-chemical feedstocks. Substitution of alternative fuels for those uses will be very difficult. A 25% reduction in domestic petroleum use by the year 2000 would seem to be a very ambitious target, but will be assumed in the example. This corresponds to a 10 quad reduction in oil demands relative to a 40 quad oil consumption and an 80 quad overall domestic energy consumption expected in 1980.

With some escalation in natural gas prices, hopefully the supply of this fuel resource can be maintained at the current level of about 20 quads per year for the next 20 years. The 10 quad reduction in oil consumption, then, would have to be replaced by the greater use of coal, nuclear and solar energy resources, and/or the overall reduction of domestic energy consumption.

The incremental energy contribution from nuclear re-sources over the next 20 years can be estimated with some confidence on the basis of existing commitments for nuclear power plants. The installed U.S. nuclear capacity in 1979 was about 50 GWe and another 100 to 150 GWe is under construction or planned. The input energy for 50 GWe of nuclear reactors is approximately 3 quads per year. Assuming, somewhat realistically, the culmination of another 120 GWe by 2000, the incremental energy input from nuclear energy would be 7 quads at that time.

Contributions from solar energy in the form of hydro-electric power, wind power, woodburning, etc., accounts for about 4 quads of equivalent energy input at this time. As pointed out by the CONAES report, [8] it is very unlikely this will increase substantially by the year 2000. Here it is assumed, somewhat optimistically, that another 3 quads might come from solar energy by 2000, increasing its yield to around 7 quads or almost 10% of the current energy consumption in the U.S. Hence, the sum of nuclear and solar additions might contribute an incremental 10 quads per year by 2000, i.e., equivalent to the decrease in oil consumption. Clearly, there is considerable uncertainty in both the energy savings from reduced oil usage and the supplemental energy available from nuclear and solar energy, thereby making the calculation of coal makeup only approximate.

With some 80 quads coming from identified resources, the deficit between actual energy consumption in the year 2000 and the identified 80 quads would presumably require the increased supply of coal. The impact of this is shown in figure 2.3 where the required growth rate for the domestic coal industry is shown as a function of the projected total-energy consumption in 2000. Hence, if zero energy growth occurs between 1980 and 2000, the required growth rate for coal production would be zero over that period. If domestic energy consumption grows to 115 quads per year by 2000, a consumption consistent with some forecasts, coal supply would have to grow at a rate of almost 6% per year, i.e., approximately a trebling over the 20-year period. Such a growth rate would stretch both coal-mining and transportation capabilities to the limit.

Probably a more plausible upper limit to U.S. energy consumption in 2000 would be 100 quads/year, i.e., an increase of about 1 quad/year for the next 20 years consistent with trends in the 1970s. That level of energy consumption in 2000 would require a 4% per annum growth of coal--still a very impressive growth requirement. Probably, then, a lower level of 80 quads and an upper level of 100 quads might bracket the plausible range of U.S. energy consumption levels one might expect in 2000.

At a consumption level of 74 quads in 2000, as implied by the cycle-adjusted-logistic projection, the U.S. energy consumption problem would virtually disappear. That, however, would be a most imprudent target for national planning.

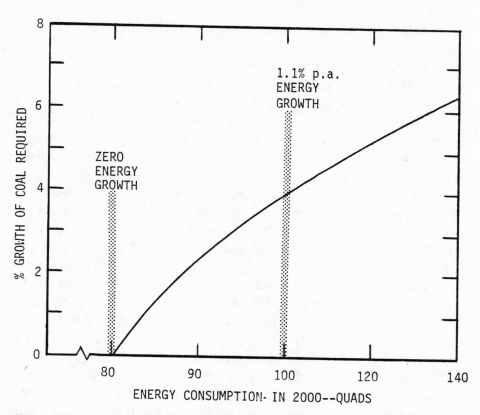

Figure 2.3 Required U.S. growth rate of coal consumption as affected by energy consumption level in 2000

Moreover, if an energy consumption level of around 74 quads should occur in 2000, it would portend an energy growth problem of enormous proportions for the period shortly after 2000. That potential problem will be examined subsequently.

This illustrative approach for indicating the magnitude of the supply/demand problem for alternative fuel resources is convenient because the statistic having the largest uncertainty, viz., total energy consumption 20 years hence, can be identified as a parameter. The supply problem for the alternative fuel, then, can be related to any plausible projection for total energy. The same approach can be used for world energy growth, to put that problem in perspective relative to the U.S. problem.

The world consumption of total energy is expected to be around 300 quads in 1980. Of that, some 130 quads is expected to be supplied by petroleum and another 60 quads from natural gas. It is most unlikely that worldwide petroleum use can be reduced significantly during the next 20 years. Yet, it is equally unlikely that worldwide supply of oil will increase significantly. Likewise, the supply and demand of natural gas will probably not change significantly. Hence, it will be assumed here that the world supply and demand of petroleum and natural gas resources will remain at about 190 quads, barring the possibility of a major political disruption.

The worldwide installed nuclear capacity is somewhat more difficult to estimate, but the uncertainty of the projection is not critically important relative to the qualitative nature of this examination. Likewise, the worldwide solar energy projections have considerable uncertainty, but are not particularly crucial in this illustration. It will be assumed that the world installed nuclear capacity will reach 1000 GWe by 2000, a projection that is somewhat lower than most projections, but consistent with lower growth planning now occurring. The increase in capacity over 1980 would be approximately 850 GWe on this basis. That increment of nuclear capacity would contribute an additional 50 quads of input energy by 2000. And the incremental contribution of solar energy is assumed to be 10 quads. Hence, some 360 quads of energy input are identified from resources other than coal.

Figure 2.4 shows the requirement for the growth of coal resources internationally to fill the supply/demand gap, assuming the deficit is met only by coal growth. In this case, world energy consumption would have to be limited to 360 quads per year to avoid any growth of coal supply. At a consumption of 500 quads in 2000, a total energy generally

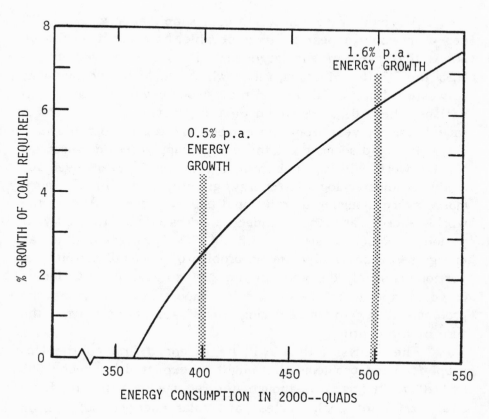

Figure 2.4   Required world growth rate of coal consumption as
affected by energy consumption level in 2000

projected for the world, coal production would have to increase at a rate of about 6% per year. Even assuming a
cycle-adjusted-logistic growth, world energy consumption
would reach about 400 quads in 2000 and coal production
would have to grow at a rate of about 2% per year (in
contrast to less than zero for the U.S. case). Probably, then,
the 400 to 500 quad energy bracket for the world is roughly
consistent with the 80 to 100 quad bracket for the U.S.

A growth rate for world coal supply approximating 2% per year would appear to be achievable, however it should be noted that most of the large coal fields are located in the U.S., the USSR and China, all of which would be inconvenient to western Europe, Japan and the various developing industrial nations where demands would be the greatest. But a growth rate of 6% per year would be extremely difficult to achieve.

It would seem, then, the energy supply/demand problem for the world will be significant in the next 20 years even for a cycle-adjusted-logistic energy growth, and will be much more severe assuming a traditional energy growth. And, if the world energy growth is indeed moderated in the 20-year period to 2000, consistent with the CAL projection, then an energy resource supply/demand problem of almost disastrous proportions is simply delayed for the 2000 to 2025 period. Hence, it would appear to be even more important to examine the problems likely to arise during the 25-year period after the turn of the century.

The analysis, then, will be extended first to the U.S. supply/demand problems that might be expected between 2000 and 2025. In this case, though, the required growth rate for a *combination* of coal, nuclear and solar energy input will be calculated for a range of total energy consumptions in the year 2025. Moreover, the growth rates are calculated for two separate assumptions on the energy demand in 2000, since that value cannot be forecast uniquely.

It is assumed that oil and gas account for 50 quads in 2000, but only 30 quads in 2025. That assumption may actually be optimistic, since it presumes the supplies of oil and gas in 2025 are 50% as large as the supplies in 1980. In all cases, then, the gap between energy demand and the oil/gas supply is assumed to be filled by some combination of coal, nuclear and solar resources.

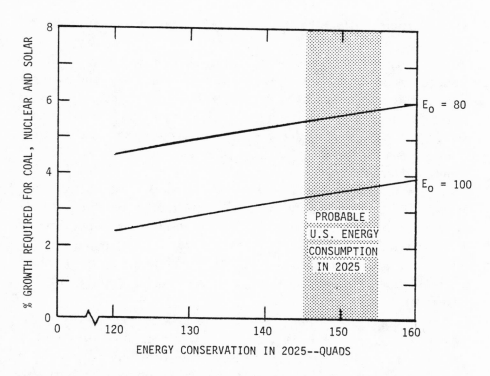

Figure 2.5   Required U.S. growth rate of coal, nuclear and solar
             energies as affected by energy consumption level in
             2025 (for two consumption levels in 2000)

        The results of these calculations are summarized in
figure 2.5.   If a zero energy growth can be achieved in the
U.S. between 1980 and 2000, a combined alternative-energy
growth of 5 to 6% per year will be required to reach 130 to
160 quads in 2025. Recognizing that growths this large will
be impossible for one or two of the alternatives, this means
the growth requirements for the other one or two will be even
greater. If the growth begins from a level of 100 quads in 2000,
the required growth rates for the alternative resources are reduced to 3 to 4% per year, a-not-so-insignificant growth. One
must conclude, then, that the energy supply/demand problem in

the U.S. will probably be more serious after the turn of the century.

But, the world energy supply/demand problem will be staggering at that time. Again, assuming that oil and natural gas supplies decrease by a factor of two, the combined growth rates imposed on coal, nuclear and solar resources are shown by figure 2.6. If a moderated energy growth, consistent with the cycle-adjusted-logistic growth, is assumed prior to 2000, a combined growth rate of 8 to 9% per year would be required for the combination of alternative resources, with even heavier burdens probably required on one or two. Such a rate of growth would appear to be impossible without major technological changes. Even assuming a continuing strong energy growth from 1980 to 2000, a combined growth rate of 5 to 6% per year would be imposed on world supplies for the alternative energy technologies.

In summary, then, the supply/demand problem for oil resources both in the U.S. and the world is already critical and will become even more serious over the long range. The requirement on coal to bridge the energy-supply gap between 1980 and 2000 appears to be substantial assuming traditional energy-growth projections, but only modest assuming a cycle-adjusted-logistic growth. Indeed, the CAL projection for U.S. growth would suggest no problem.

But, energy supply problems in the period 2000 to 2025 appear to be formidable under almost any circumstances. And the problems are much more serious worldwide than for the U.S. As one more indication of the world problem, the shipping tonnage for coal in the year 2025 is estimated to be four to five times that of oil in 1980, even assuming that coal only accounts for 50% of the energy growth between now and 2025. While coal must remain as one of the possible solutions to the evolving world energy crisis, the merits of nuclear resources are apparent.

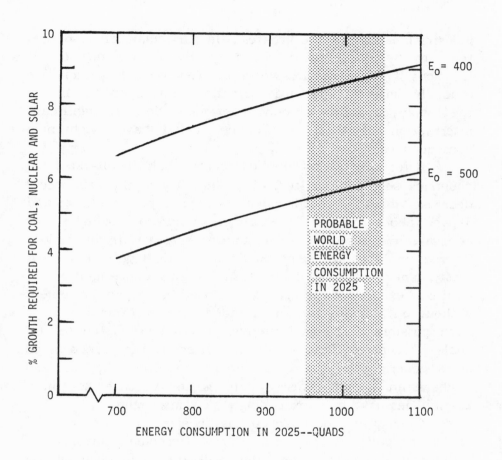

Figure 2.6   Required world growth rate of coal, nuclear and solar
             energies as affected by energy consumption level in
             2025 (for two consumption levels in 2000)

ENERGY-DEMAND IMPLICATIONS ON ENERGY INDEPENDENCE

While overall energy supply/demand problems may be most
severe beginning some 25 years from now, the petroleum
supply/demand problems may be most threatening to world
stability in the next 10 to 20 years. Nature has not been
particularly kind in distributing fossil-fuel resources where

industrial needs are the greatest. In particular, oil resources tend to be concentrated in a relatively few number of locations. As a result, world oil requirements--which are crucially important for transportation fuels--lean heavily on the supply capability of only a few countries. A significant interruption of those oil supplies could have frightening consequences.

If for example, oil supplies from only half the oil-exporting countries should be halted, oil availability in the U.S. would decrease about 25%, but in most western European countries and Japan, about 50%. The struggle between countries for adequate oil supplies would undoubtedly result in serious international strains. And public or political pressures in at least some countries could lead to the contemplation of military excursions to re-open closed oil-supply channels. Without oil supplies for an extended military excursion, it is even possible that some countries could be tempted to use nuclear weapons. Those grim possibilities are being recognized more generally as the result of recent events in Iran and Afghanistan. But, energy policies in response to those possibilities have been exceedingly slow in coming.

Clearly, oil dependence has become an internationally explosive political problem. Even Herculean efforts on greater coal, nuclear and solar utilization can probably do very little to minimize that threat in the near term. Strong conservation programs, maximum use of alternative energy sources (where substitution is possible) and petroleum storage could, possibly, help somewhat, but the world will undoubtedly have to pursue energy policies for one to two decades that will require enormous patience and some altruism--character-istics not generally found at national levels.

Figure 2.7 shows the U.S. history of domestic and imported oil utilization, with extrapolations of consumption rates to the end of the century. Bands are shown indicating

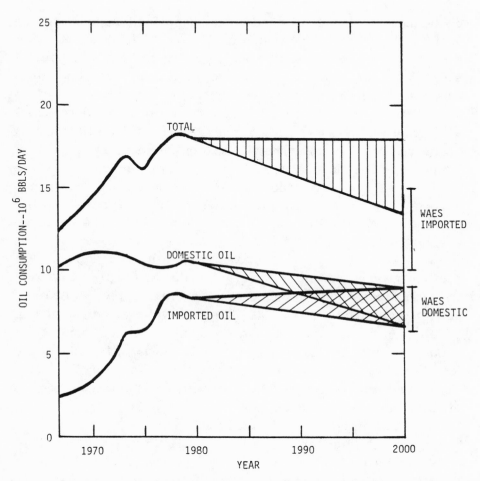

Figure 2.7   History and projections of domestic and imported oil
utilization in the U.S.

the WAES demand projections for the year 2000. As previously
indicated, a lower consumption rate in the U.S. should be possible and policy planning should obviously pursue that objective.
Trends following the 1973-74 embargo, however, have not
been encouraging in the absence of stronger national policies.
Prior to the embargo, domestic oil consumption was at a level

equivalent to almost 35 quads, with import oil accounting for about 37% of our oil consumption.[14]   In 1975,   just following the embargo, a sensitized country limited their appetite to about 33 quads with imports again accounting for about 38%. However, in 1976, consumption grew to 35 quads, in 1977 to 37 quads, and in 1978 to 38 quads, for an average increase of 5% per year between 1975 and 1978. In 1978, oil imports accounted for 45% of domestic oil consumption, in spite of a temporary relief in 1977 from the Alaskan pipeline.

A very important consequence of the oil-import problem, beyond that of possible international confrontations, is the trade deficit associated with imported oil. Indeed, because of the higher cost of oil per energy content, relative to other fuels, the economic implications of oil dependence are particularly significant.   This can be illustrated by comparing the annual cost of fuel resources for the supply of heat to generate electricity from 100 GWe of power stations (assuming a 70% load factor for the plants). This particular capacity is about one-sixth the capacity of all U.S. stations, but about the installed capacity one might expect in the year 2000 for Japan and the larger European countries. The comparison of oil, coal and uranium is as follows:

- $30 billion for oil (at $30 per bbl),
- $ 6 billion for coal (at $1 per 10^6 BTU), and
- $1.4 billion for U_3O_8 (at $43 per lb).*

Even with enrichment, the cost of uranium fuel would be less than $3 billion.   Clearly, countries interested in minimizing their trade deficits for fuel resources should presumably transfer their dependence from fossil fuels to nuclear fuels as expeditiously as possible.   That policy also has the advantage that fuel storage for future contingencies is more practical.

*Assuming a once-through cycle in the LWR.

Though only half the U.S. oil supplies are imported, the financial burden on our trade balance has been substantial. In 1975, the U.S. net export profit, not including oil imports, was $46 billion. With an oil import cost of $35 billion, the country was still able to show an overall export profit of $11 billion. In 1977, the U.S. net export profit, excluding oil, was $17 billion.   With an oil import cost of $44 billion, the U.S. showed a trade deficit of $27 billion. Assuming no reductions in oil imports by 1985, i.e., about 8 million barrels per day (or 3 billion barrels per year) and a cost of, say $30 per barrel in 1979 dollars, the oil import cost in 1985 could reach almost $100 billion in constant dollars. If the price of imported oil escalates at a value of 5% per year above normal inflation, a distinct possibility in a tight world market, the 1985 import cost for oil could exceed $120 billion in constant dollars, or about $500 per person in the U.S.

Obviously, ways must be found to substitute other fuels for oil.   To explore that possibility, it is useful to examine how oil is currently being used in the U.S. Following is the distribution of oil uses that occurred in 1975: [14]

Transportation fuel:	55%
Industrial non-fuel uses:	10%
Industrial fuel uses:	11%
Residential and commercial fuel uses:	14%
Electricity fuel use:	10%
	100%

While the first two items account for two-thirds of the domestic oil consumption, these are the most legitimate and probably the least substitutable uses. An Office of Technology Assessment study of the Automobile Transportation System[15] projects that automobile fuel consumption will reach a peak before 1985 and will remain constant between 1985 and 2000 as gains in fuel efficiency compensate for added vehicle miles traveled. Deregulated gasoline prices, added gasoline tax, further improvements in fuel economy, more emphasis on mass transport and the partial use of methanol or ethanol could reduce gasoline dependence in the near term. Some slowdown in the U.S. economy could also have an impact--though that solution cannot be regarded as an attractive one. When the requirements of non-auto transportation (buses, trucks, airplanes, etc.) and non-fuel industrial uses (petrochemical feedstocks) are included. It is difficult to see how an improvement of more than 10% might be achieved in the first two sectors.

The last three sectors all use oil as a fuel for heating. Coal and nuclear energy would each be economically preferable to oil for electricity generation, though some parts of the country require the use of oil fuel because effluents from coal plants cannot be tolerated and nuclear energy has been resisted. Hopefully, those limitations ultimately can be removed.

Probably one of the most promising areas for oil substitution is the large-scale application of solar heat for space heating in homes and commercial buildings. This application of solar energy is likely to be economically attractive and reasonably effective in the next 10 to 20 years. It is possible, then, that some significant fraction of the 25% use for residential, commercial and industrial heating could be transferred to solar energy. But, from a practical point of view, it must be recognized that even a 50% substitution (which is probably not likely in 20 years) would only reduce oil consumption by 12% and overall energy consumption by 6%. With possible solar contributions to methanol production, hydroelectric generation and, perhaps, other applications, it is possible, though, that solar energy could make a 10% contribution 20 years hence.

In the next two decades, then, it might be possible to reduce oil consumption by 15% in the first two uses (65% of oil consumption) and, perhaps, 40% in the last three (35% of oil consumption) for an average reduction in oil consumption of about 25%. With domestic production held at current levels, this could allow a 50% reduction in dependence on import oil requirements--a very impressive potential.

But, oil dependence in the longer range is probably even more important. Relief by the year 2000, for example, might come from one or more of several new technology directions, including:

  - recovery of oil from shale oil or tar sands,
  - production of synfuel from coal,
  - production of alcohol from biomass,
  - use of electricity for vehicles, and
  - production and storage of hydrogen.

All of these technologies would require significantly more energy than recovery and refining of petroleum. Moreover, some of them, particularly the first two, could present new environmental problems.

The intent here is not to assess advantages and disadvantages of new technology directions, but simply to emphasize that substitutions must be found for the utilization of petroleum fuels in particular, and fossil fuels in general. Recalling the enormous energy growths projected after the year 2000, serious international problems can evolve if the world does not find ways to allow continuing energy growth, particularly outside the U.S. The passive solution of energy conservation is exceedingly important for the U.S. in the next 20 years, but more than passive solutions will be required for the world. The U.S. should not discourage active energy-technology solutions in the rest of the world.

ENERGY–DEMAND IMPLICATIONS ON ENVIRONMENTAL IMPACT

The environmental impact of both fossil fuels and nuclear fuels will be examined only from the point of view that these impacts could limit the ultimate growth of energy production using fossil fuels.   The potential environmental problems associated with fossil fuels will be summarized here, while the subject of nuclear energy wastes will be reserved for Chapter III.

One approach to classifying the environmental impacts associated with the use of fossil fuels is to identify the sphere of influence resulting from alternative uses.   Hence, the consumption of liquid fuels in densely-populated urban areas is usually most responsible for *local* atmospheric polution involving primarily hydrocarbons, nitrogen oxides and particulates.   The consumption of coal in electricity-generating stations and industrial plants, usually away from urban centers, can lead to *regional* atmospheric problems involving sulfur products and acid rains.   The consumption of all fossil fuels ultimately poses a *global* problem of CO_2 buildup in the atmosphere with its potential effect on world climates. Each of these problems will be discussed very briefly.   In general, recognition of the problems has brought local and national controls that promise to limit the severity of the first two problems, albeit at some expense and inconvenience to industry and the public. In the long range, though, the CO_2 buildup problem may be the most serious and least amenable to a solution--other than the curtailment of energy production involving the burning of carbon and hydrocarbons.

Looking first at the localized urban problems, the use of automobiles is clearly an important contributor to air pollution problems.   The OTA report on the Automobile Transportation System [15] points out that 95 million automobiles in the U.S. generated more than 81 million tons of atmospheric pollutants in 1975.   The air quality standards for either carbon monoxide or photochemical oxidants were exceeded in about one-third of the 247 Air Quality Control Regions (AQCRs) in the U.S. during that year. It was estimated that 125 to 150 million people were potentially exposed at least once during the year to concentrations of these pollutants that exceeded federal standards. Both hydrocarbons and nitrogen oxides lead to the production of photochemical oxidants, such as ozone, that are responsible for eye irritations and pulmonary problems. Carbon monoxide is a source of problems associated with the cardiovascular system. While definite statistics are unavailable on the consequences of automobile emissions, it is estimated that as many as 4000 deaths and four million illness-restricted days per year can be attributed to this source.

Atmospheric pollution has shown some improvement in the last few years due to controls imposed by the 1970 Clean Air Act and the 1977 amendments. Projections of regional air quality for the year 2000 indicate that violations of the standards may still occur in at least 25% of the AQCRs. In the very long range, only the use of electricity for transport or the evolution of a hydrogen economy would seem to be assured solutions to the urban pollution problems. Either of these directions would involve energy investments for the production of the consumable energy resources, thereby increasing effluent problems associated with industrial power plants.

Sulfur dioxide and particulate sulfur-related emissions from coal-burning electricity-generating stations and industrial process-heat plants can contribute importantly to the local atmospheric problems. In addition, though, it has been recently discovered that SO_2, sulphate and NO_x contaminants can have a significant impact at much more remote locations, i.e., at distances several hundred miles away. These problems were already recognized and discussed, for example, in a National Academy of Sciences report in 1975 and in the MITRE/Ford Foundation report published in 1977. [16]

With the use of coal containing 3% sulfur, the National Academy of Sciences report suggests that 2 to 100 deaths per year could result from a 1000 MWe power plant. If half the generating capacity of the U.S., i.e., around 300 GWe, should utilize this type of coal, the implied death rate would be 600 to 30,000 deaths annually. With the use of flue-gas scrubbing equipment, the sulphur emissions are reduced by an order of magnitude; and with a combination of low-sulfur coal and scrubbers, the emissions are reduced some 50 times. In this case, the implied death rate would be 12 to 600 per year. However, as sulfur emissions are reduced, the importance of other pollutants, such as the nitrogen oxides, become relatively more important.

Again, though, much progress has been made since the Clean air Act of 1970 was enacted. Sulfur releases have been steadily reduced, although the reductions have come at considerable expense to the ultimate customers of electricity.

Though the direct health effects of sulphur-related pollutants have tended to get primary attention, the effects of acid rain resulting from the atmospheric scrubbing of sulfur and nitrogen oxides are only beginning to get more recognition. The use of tall exhaust stacks to reduce the local impact of these effluents has had the effect of lifting the effluents to higher atmospheric strata for better dispersal. The conversion of SO_2, particulate sulfates and NO_x to acids results ultimately in rainfall at remote locations having an abnormally high acidity. This subject has recently been getting significant attention in the more sophisticated journals. [17, 18, 19]

The increase in acidity of rainfall has been observed to be one to two orders of magnitude higher than normal in the Northeastern United States and in the Scandanavian countries of Europe. This acid rainfall is already seriously affecting the propagation of fish in fresh water streams and is apparently also retarding the growth of trees. Detrimental effects on both metallic and concrete structures are also becoming more prevalent.   While the use of still more efficient scrubbers to clean smoke-stack   effluents could, of course, improve the problem, the increased use of fossil fuels in the next few decades will probably make it difficult to offset potential improvements from better scrubber technology.

While the local and regional atmospheric pollution problems previously discussed are amenable to at least partial solutions, the buildup of CO_2 in the atmosphere on a global scale is a much more difficult problem.   There has been increased speculation in the scientific community about the long-range impact of the growing CO_2 atmospheric content on the climate of the world. In 1977, a panel of the National Academy of Sciences issued a warning that the growing consumption of fossil fuels may have drastic effects on the world's climate over the next century or two.   A recent study by Oak Ridge scientists [20] has shown that the CO_2 content of the atmosphere has been increasing at a rate of about 4% per year in the last two decades. It is their projection that continuing increases will show significant serious effects in our weather patterns in less than 100 years. This problem can be resolved in the long range only by a transfer to nuclear and solar energy with hydrogen replacing carbon and hydrocarbon fuels for combustion.

SUMMARY

In summary, then, both limitations of fossil fuels and their growing environmental impacts will force the end of the fossil-fuel era. Active national and international policies will be required to move the U.S. and the world through the very difficult transition to new energy sources. A passive policy of energy conservation has some merit for the U.S., at least for the next 20 years.   Moreover, the increased use of solar energy during this time can be beneficial and can allow a better assessment of solar-technology potential for the future.

However, the continuing rapid growth of energy consumption in other parts of the world, and the probable new surge of energy growth in the U.S. and the world beginning around the year 2000 make it imperative that solutions for the intermediate and longer-range future be more than passive ones. The world will be moving through a delicate period, both relative to energy supply/demand balances and environmental impacts, during the next few decades. Technology preparation and policy planning for these delicate times are exceedingly crucial.

REFERENCES

1.   Schlesinger, James R., "Energy Risks and Energy Futures", Wall Street Journal, August 23, 1979.

2.   Netschert, Dr. Bruce C., "The International Energy Situation: Outlook to 1985", Electric Perspectives, 77/4, pp. 18-25.

3.   Morgenthaler, Eric, "Soviet-Bloc Countries Also Face Problems Meeting Energy Needs", Wall Street Journal, July 6, 1979.

4.   "Energy Global Prospects 1985-2000", Report of the Workshop on Alternative Energy Strategies (WAES), McGraw-Hill, 1977.

     -----"Energy Supply-Demand Integration to the Year 2000", (WAES), MIT Press, 1979.

5.   Stobaugh, R., and Yergin, Daniel, etc., "Energy Future", Random House, New York, 1979.

6.   Olds, F.C., "Outlook for Breeder Reactors", Power Engineering, March 1979.

7.   Imai, Ryukichi, "What Will Be The Impact of INFCE?", Nuclear Energy International, December 1979, pp. 66-67.

8.   "Energy in Transition, 1985-2010", Final Report, Nuclear and Alternative Energy Systems, National Academy of Sciences, Washington, D.C., 1979.

     -----"Alternative Energy Demand Futures to 2010", National Academy of Sciences, 1979.

9.   "The Need for and Deployment of Inexhaustible Energy Resource Technologies", Report of Technology Study Panel, U.S. Energy Research and Development Administration, September 1977.

10. "Annual Report to Congress 1978", Energy Information Administration, DOE/EIA-0173/3.

11.   Giraud, Andre, "World Energy Resources", presented at Conference on World Nuclear Power, Washington, D.C., November 1976.

12.   Häefele, Wolf, "Global Perspectives and Options for Long-Range Energy Strategies", Energy, Volume 4, pp. 745-760, January 1979.

13.   Kiely, John R., "Energy-Looking Into the 21st Century", presented to the American Power Conference IEEE Luncheon, Chicago, Illinois, April 1979.

14.   Monthly Energy Review, January 1980, DOE/EIA-0035/01-(80), U.S. Department of Energy.

15.   "Technology Assessment of Changes in the Future Use and Characteristics of the Automobile Transportation System", Summary and Findings, Office of Technology Assessment, Washington, D.C., 40-964 O-79-2.

16.   "Nuclear Power Issues and Choices", Sponsored by the Ford Foundation, Administered by the MITRE Corporation, Ballinger Publishing Company, 1977.

17.   Likens, G.E., Wright, R.F., Galloway, J.N. and Butler, T.P.J., "Acid Rain", Scientific American, Volume 241, No. 4, October 1979.

18.   Alexander, T., "New Fears Surrounding the Shift to Coal", Fortune Magazine, November 1978, pp. 50-60.

19.   Rosenfeld, A., "Forecast: Poisonous Rain", Saturday Review, September 1978, pp. 16-18.

20.   Baes, C.F., Olson, J.S., and Rotty, R.M., "Carbon Dioxide and Climate: The Uncontrolled Experiment", American Scientist, May-June 1977, pp. 310-320.

Chapter III

Nuclear Energy Directions:
The Changing Issues

Early in 1977, the very influential report "Nuclear Power Issues and Choices" [1] was published. The policy analysis summarized in that report was the result of a year-long study by a Nuclear Energy Policy Study Group, administered by the MITRE Corporation and supported by a Ford Foundation grant. The report was a significant one primarily because the recommendations of the report, rightly or wrongly, have become the basis for subsequent government policies.

The conclusions of the report might be summarized as follows:

    (a)   Ample U.S. uranium resources should be available to satisfy nuclear needs for several decades without plutonium recycle, particularly in view of the newly-projected growth for nuclear energy;

    (b)   The economics of more resource-efficient nuclear technologies, such as plutonium fuel recycle and breeder reactors, are only marginally attractive at best;

    (c)   A continuing U.S. commitment to the implementation of these technologies could encourage other countries to follow that policy, thereby increasing the availability of weapons-usable plutonium throughout the world; and, conversely,

    (d)   A U.S. decision to forego those technologies could encourage non-weapons countries not to acquire facilities and materials that might subsequently be used for military purposes.

In general, those conclusions were probably valid in the short term, though critics from the nuclear establishment, both in this country and abroad, severely attacked the report.

The profusion of criticism that arose from the nuclear community might similarily be summarized by the following:

(a) The argument of uranium-resource exhaustion was resolutely defended, largely on the grounds that government resource projections were optimistic;

(b) Large increases in reprocessing and refabrication costs for recycle fuels and the escalation of capital cost estimates for breeder reactors were generally discounted or ignored; and

(c) The continuing development and deployment of fast breeder reactors in the U.S. was endorsed unequivocally.

Particularly in retrospect, the industry response was obviously an overreaction to the Ford Foundation/MITRE report.

Some three years later, it is more apparent that the economic and resource bases for the Ford Foundation/MITRE conclusions were probably correct, at least in the U.S.   However, the report could be faulted for two reasons:

- the conclusions were based primarily on U.S. conditions and did not adequately take into consideration the concerns of energy growth and energy dependence in other parts of the world, and

- the recommendations focused primarily on passive solutions rather than looking for active solutions to the world problems.

The objections of the nuclear community could also be faulted for at least two reasons, viz:

- there was an overriding inclination within the nuclear community to ignore the changing issues that already

were becoming apparent relative to economics and
resource availability, and
- there was an equally strong resistance to consider
future changing issues that in any way could suggest
a deviation from the traditional strategies.

It is the intent of this chapter to examine the changing
issues as they relate to nuclear energy. As will be seen, the
changing issues, which include economic, societal and technology
changes, can have profound repercussions on both government
and industry planning. A greater awareness of those changing
issues might, hopefully, provide a better perspective for
technology and institutional redirections to be suggested in
subsequent chapters.

THE CHANGING ISSUES
The changing issues refer not only to conditions that have
changed, but also to those that might be expected to change
in the future. The changing issues will be classified under the
general subjects of:
- nuclear growth,
- uranium-resource availability,
- oil-resource availability and nuclear substitutability,
- nuclear economics,
- environmental aspects of nuclear energy, and
- industry and government institutions.

An overview of the basic issues will be presented briefly, with
more detailed discussions following.
Perhaps the most important changing issue is that of
the lower energy growth now being projected. We can
probably all be excused for overlooking this possibility in the
days when we were conditioned to cheap energy and during an
economic period when growth was abnormal. The situation has
clearly changed, however, in the last few years. To put this

problem in perspective, only five years ago we were projecting an installed nuclear capacity of 1200 GWe in the year 2000 and some 4000 GWe by the year 2020 in the U.S. [2] Today our projection is less than 300 GWe in the year 2000 and perhaps 700 GWe by the year 2020. This is a reduction of more than five times.

While there may be valid reasons why a rapid nuclear growth may be unnecessary in the U.S. for the next 20 years, there are probably equally valid reasons why nuclear growth may be more necessary in other parts of the industrial world. As indicated in the previous chapter, energy growth in other parts of the world is generally on the more rapidly ascending portion of a typical growth curve. And energy growth cycles, if they are worldwide, should have a smaller effect on the steeper world growth than on the already-saturating U.S. growth. Moreover, most of the other industrialized nations are much more dependent on externally-supplied energy resources and can be expected to correct that problem more aggressively than will the U.S. Just how nuclear growth might evolve in the U.S. and the world will be examined in some detail subsequently.

A consequence of the changing energy growth in the U.S. is the severe effect an extended low growth could have on the nuclear industry. If the growth is destined to be low for 10 to 20 years, as the energy-cycle pattern might suggest, this could seriously threaten the strength of a nuclear industry already suffering from growth pains. And, if another large growth can indeed be expected some 20 years from now, it will be crucially important to preserve and build on our present industry strength to meet the demands of that growth surge. Related to this possibility is the problem of supporting continuing technology development to prepare for such a subsequent energy growth surge. Almost certainly, a weakened nuclear industry cannot afford to invest heavily in R&D during an abnormally slack business period.

The second changing issue is the obsolescence of the uranium resource exhaustion argument--at least for the next 30 to 50 years. Based on the nuclear growths that were projected in the early 1970s and the uncertainty of projected U.S. uranium resources, a legitimate concern developed on the possible early depletion of high-grade uranium resources. This strongly suggested an urgency for the prompt development and deployment of fast breeder reactors that could generate more nuclear fuel than that consumed.

As a result of the National Uranium Resource Evaluation (NURE) program being sponsored by DOE, U_3O_8 reserves and resources up to forward costs of $50 per lb have been classified more precisely than before. The projected reserves and resources most recently reported by DOE [3] for forward costs up to $50 per lb, are 4.1 million ST (short tons) U_3O_8. The Ford Foundation/MITRE report argued that resource projections by the government were very pessimistic and with continuing effort a much larger quantity of uranium would probably be found. On the other hand, a recent CONAES study [4] by the National Academies of Science and Engineering argued that the DOE results were probably optimistic and a more prudent resource estimate might be approximately 2.0 million ST U_3O_8. A review of the entire uranium resource problem was recently completed by Stoller Associates [5] under a contract to DOE. That very thorough study concluded that the CONAES results were probably pessimistic, the Ford/MITRE results were probably optimistic and that the DOE projections are probably the most reliable.

While it is likely that the threat of resource exhaustion has been postponed at least some 40 years by the lower U.S. growth projections, a subsequent surge in nuclear growth could, nevertheless, put heavy pressures on the *rate* of mining and milling uranium resources. That problem of production rate is somewhat different than the more traditional concern of resource exhaustion, and will get more attention in a following section of this chapter.

A changing issue closely related to the uranium resource question is the probable introduction of Advanced Isotope Separation Technology (AIST) within the next 20 to 30 years. This technology could have a very important effect on nuclear energy in at least two respects. First, the successful development and deployment of AIST would probably reduce substantially the amount of fissile uranium normally lost in the enrichment process. The use of AIST could, then, have the effect of stretching U_3O_8 supplies by 20 to 30%. Secondly, the lower capital costs and energy costs for an AIST plant could reduce the enrichment cost component of the nuclear fuel cycle cost. With a lower cost for separative work, a higher price could actually be afforded for U_3O_8.

In contrast to the decreasing concern related to uranium supply adequacy, the concern of oil-supply adequacy has reached a crisis level as indicated in the previous chapter. A probable changing issue, then, is the increased attention that will be directed toward possible nuclear-energy substitutions for oil uses, over the next few decades--assuming the problems of nuclear energy can be solved, particularly the public acceptance problem. Since nuclear energy technology has traditionally been dismissed as a candidate for solving the transportation fuel supply problem, this changing issue has yet to be manifested. Nevertheless, nuclear energy can conceivably make a significant contribution even to the solution of the portable-fuel problem, either through high-temperature process heat for the synthesis of fluid fuels or through vehicle propulsion systems that use electricity generated from nuclear plants.

A national response in this critically important area has been surprisingly slow in view of the serious international consequences of failing to solve the problem. In fact, the one reactor technology that can supply high-temperature process-heat energy, the HTGR (high temperature gas-cooled reactor) is supported at a relatively low level of federal funding and, indeed, has barely survived several attempts to cancel it completely.

If and when a greater role is assigned to nuclear energy to contribute to the resolution of the oil-supply problems, it is possible that the higher-grade uranium resources may be depleted somewhat more rapidly and the need for breeder reactors may be accelerated. Nevertheless, it would seem appropriate to assign first priority at this time to the very pressing problem of relieving pressures on oil supply, recognizing that the longer-range problem of a uranium supply problem should continue to receive development and demonstration attention.

The changing issue of nuclear energy economics is also an important one. As a result of recent experiences in building reprocessing plants, both in this country and abroad, it has become apparent that the unit cost for reprocessing spent fuel will be about an order of magnitude greater than that expected some ten years ago. In addition, there is mounting evidence that the cost penalty of refabricating radioactive bred fuels will be substantially higher than previously assumed. While plutonium recycle was clearly accepted as economically advantageous ten years ago, more recent studies suggest that the economic advantage has seriously deteriorated. Large increases in U_3O_8 prices, could of course, improve the economic incentive for recycle, but lower separative-work costs could have the opposite effect. Because of these changing economics, the whole strategy of plutonium utilization must be re-examined, quite independent of weapons-proliferation concerns. This is not to imply that the concept of plutonium utilization should be abandoned, but rather that more economic utilization strategies should be sought.

The effect of higher recycle costs also critically affects the attractiveness of near-breeder and breeder reactor concepts. In addition, the higher capital cost estimates now being suggested for the Liquid Metal Fast Breeder Reactor (LMFBR) are even more disturbing. Some five years ago, it was expected that the capital cost of the LMFBR would be about 10% greater than that of the LWR. Today, the capital cost penalty appears to be at least 25%, with some cost-penalty projections as high as 75%. If a penalty in this range persists, together with the recycle cost penalty, it will be difficult, indeed, to justify the introduction of LMFBR plants until U_3O_8 prices reach about $200 per lb. This being the case, it would seem even more necessary to examine alternative transitional technologies that could encourage the introduction of resource-efficient reactors and fuel cycles. This issue is a particularly important one since there tends to be some belief that changes in directions imply a repudiation of a previously accepted policy direction. While it is probable the long-range nuclear strategy of breeder and near-breeder reactors will not be affected, it is likely that the technology and institutional directions to accommodate the *transition* may require much more thought. That is a major theme of this book.

Appropriate world directions on breeder reactors and fuel cycle may be quite different, though, from the optimum U.S. directions, both because of differences in energy growth and differences in energy dependence. The greater urgency to achieve some measure of energy independence in other industrial nations may force those countries in the direction of nuclear fuel recycle and deployment of breeder reactors even before they might be economically optimal. Probably even more important is the difference in direction appropriate for high-energy-consuming industrialized nations and the lower-energy-consuming nations on the threshold of industrial expansion. While nuclear energy can be an important factor in the early growth of these neo-industrial states, the timing for fuel recycle and breeder reactors may be, coincidentally, more similar to the timing needs of the U.S. than the needs of the other industrialized countries. The differences, then, between the needs of various types of nations must be recognized and, where those needs differ, they must be accommodated.

The changing issues associated with the environmental aspects of nuclear energy are also exceedingly important. These issues include both the safety of reactor operations and the proper handling of nuclear fuels and their waste products. The Three Mile Island nuclear incident has certainly made the public more sensitive to nuclear safety. Presuming that the availability of nuclear energy is at least one ingredient to energy survival, particularly in the longer range, much more attention must be given to minimizing the frequency of even minor nuclear incidents. Perhaps more importantly, though, attention must be given to minimizing the *consequences* of serious incidents, should they occur even infrequently. It is likely that even a *low-probability* high-consequence nuclear accident will not be tolerable to the public. Most of the nuclear safety attention has, to date, been directed toward minimizing the probability of accidents, i.e., the assurance of "defense-in-depth". That philosophy must still be pursued, but it must be complemented by a broader philosophy emphasizing both low probability *and* low consequence. Probably the most effective way to minimize consequences is simply locating nuclear plants at greater distances from highly populated areas.

The problem of nuclear waste disposal is an equally urgent one. Apparent uncertainties in waste-disposal directions to be pursued have had a pervasive effect on public acceptance of nuclear energy in general. Actually, the much lower volume of wastes associated with nuclear energy should offer some opportunities for improvements in the environment relative to the very difficult task of containing pollutants from the continuing use of fossil fuels. But, that opportunity can be realized only if public acceptance can be assured for all aspects of nuclear energy.

In some sense, perhaps, the concern of nuclear weapons proliferation can also be classified as an environmental problem. This issue has become an extremely emotional one with a spectrum of opinions ranging from one extreme suggesting a complete moratorium on nuclear energy to the other extreme suggesting civilian nuclear energy should not be encumbered with this problem.   There is a growing feeling, with some justification, that "technology fixes" will probably be of limited value and that "institutional fixes" will be ultimately required to avoid the further proliferation of nuclear weapons. The study of these problems has seriously impaired nuclear technology development for some two years, while government studies and recommended solutions have been pursued. It is tempting to believe that a more aggressive industry role in exploring and adopting both new technological and institutional ideas might help to restore some momentum to the nuclear program.

The changing issues associated with industry and government institutions have yet to become more apparent. Already there has been some recognition that the commercialization problems associated with new technologies will be enormous.   Nuclear industry experiences within the last five years have clearly shown that development costs, introduction costs and business risks arising from complex and demanding technologies are beyond the prudent investment resources of even the most affluent industries. As was amply demonstrated with the reprocessing industry, even government policy re-directions pose an unacceptable risk to potential nuclear industry entrepreneurs.   It is a distinct possibility that the government decision to forego nuclear-fuel reprocessing in this country will make it impossible to entice U.S. industry ever again to invest in this particular type of enterprise. Hence, a clearer definition between long-range government and industry responsibilities in the nuclear energy industry will probably be required.

It would seem that the changing issues are sufficiently serious to re-examine both government and industry policies on more than a superficial basis. It would also seem useful for the industrial nuclear community to assume more initiative in understanding the most critical problems and in developing policies to respond to these changing issues in a more orderly and a bolder manner than simply defending older positions. Because of the importance of a proper recognition of the changing issues and their implications, more attention will be given in the succeeding sections to the problems involving:

- the nuclear growth issue,
- the uranium resource demand issue,
- the portable-fuels issue,
- the safety and environmental issue, and
- the institutional issue.

THE NUCLEAR GROWTH ISSUE

In the previous chapter, growth projections of *total energy* for the U.S. and the world were compared. Here, emphasis will be placed on the projection of *electrical energy* growth. The energy input to electricity generation already accounts for some 30% of the total energy consumption in the U.S. If the growths of total energy and electricity generation follow the logistic projections for the next 25 to 50 years, the fraction of domestic energy input required for electricity generation will rise to about 45% by the year 2000 and over 60% by the year 2025. The fraction of the *world* total energy consumption required for *world* electricity generation will probably be only slightly smaller than that in the U.S. Hence, the energy resources required just for electricity generation will be of paramount importance.

While it was noted previously that the growth of *total* energy could actually decrease to zero for the U.S. during the period from 1975 to 2000, the growth of electrical energy generation should still be modest during that period. Growth curves for both the U.S. and the world are projected in figure 3.1 with the dashed lines showing logistic projections and the solid lines the CAL projections. Circles are shown on the curves to indicate historical data for the last 35 years. The growth curve for the world may have been chosen slightly conservatively in view of the data for the last ten years.

From the curve it can be seen that the U.S. generated about 35% of the total world electricity in the year 1975. That fraction, though, should decrease to about 25% in the year 2000 and less than 20% by the year 2025. Once again, it is clear that world energy growth is much more significant than U.S. energy growth. As an incidental point, the cyclic variations in world electricity generation (if indeed they occur) appear to have a smaller perturbation on the world growth curve than on the U.S. growth curve.

As might be expected, the total installed generating capacity also shows a cyclic variation about a logistic growth curve. Though the deviation curve is not reproduced here, two features tend to distinguish the cyclic variation of installed capacity from that of electricity energy generation, viz:

- the amplitude of the variation tends to be
  around 30% instead of 20%, and
- the cycle peaks tend to be delayed about five
  years beyond those of the electricity generation
  cycle.

Those differences should probably not be unexpected since the time between a plant construction commitment and a plant startup is at least five to ten years and, therefore, some additional inertia is built into the generating-capacity growth. Shown in figure 3.2 are the logistic and cycle-adjusted-logistic growth curves for the installed electricity generating capacity, assuming a 30% cyclic variation about the logistic curve. Curves are shown both for the U.S. and the world.

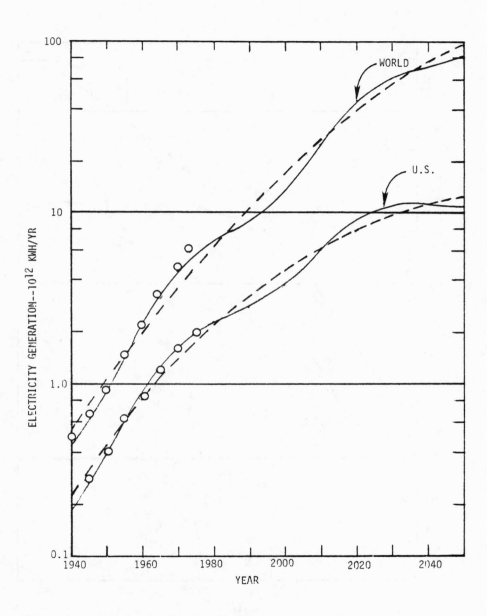

Figure 3.1 U.S. and world growth curves for electricity consumption

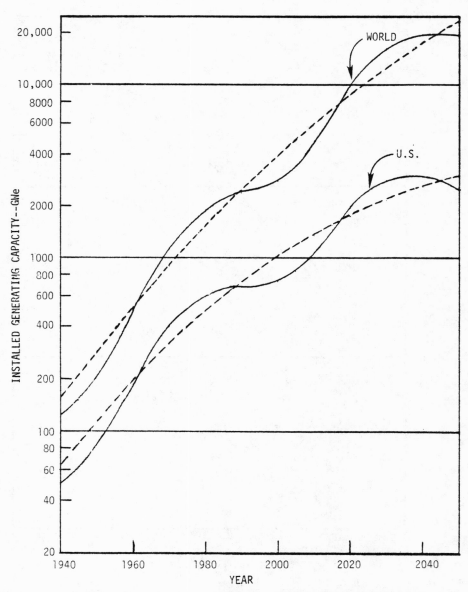

Figure 3.2   U.S. and world growth curves for installed electricity–
             generating capacity

Of greater interest, perhaps, is the number of generating plant additions from year to year. The implied addition rates are developed simply by calculating the differences between installed electrical capacity for successive years. Some particular caution must be exercised, however, in interpreting difference data too literally. Since the installed capacity is already a relatively uncertain projection, particularly with cyclic variations superposed, the successive differences between these data points can be even more uncertain. Hence, the projection of U.S. and world net addition rates is meant only to be illustrative of growth differences.

Figure 3.3 shows the net addition rates of electricity capacity for the U.S. and the world covering the period 1940 to about 2040. As might be expected, the cyclic variations in total installed capacity have a much more dramatic effect on net additions per year. While the shapes of the curves are somewhat similar, the magnitudes for U.S. and world addition rates are significantly different. For example, the curves would imply that net additions in the period between 1980 and 2000 will fall to 3 to 5 GWe/yr in the U.S., while the additions to world generating capacity will fall to 30 or 50 GWe/yr. Perhaps more important is the difference in addition rates around the year 2020. The curve suggests that net additions would reach about 90 GWe/yr in the U.S. and a remarkable 600 GWe/yr worldwide. To put this in perspective, the worldwide net additions in a single year would be equivalent to the total installed electricity generating capacity in the U.S. for the year 1978. It should also be noted that the net addition rate does not include replacement plants, which would increase the generating capacity additions per year even more.

As can be appreciated, the capital funding required for such a high electricity growth will be staggering. It is frequently noted that the financing of large capital ventures in the future may be a controlling item in economic growth. However,

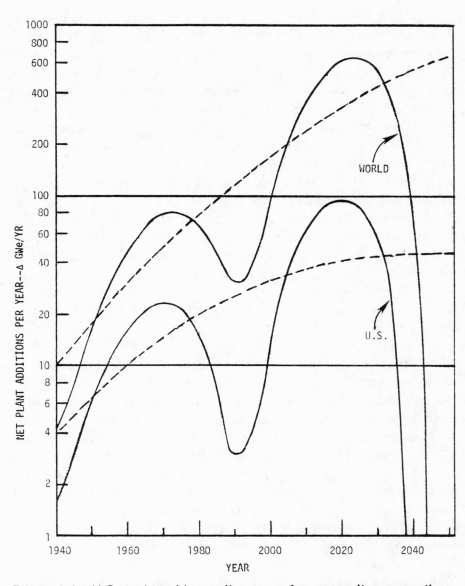

Figure 3.3   U.S. and world growth curves for generation capacity
             additions per year

we are probably currently conditioned by the financial problems imposed on us by the peaking of an economic cycle. If one accepts the concept of a long economic cycle, the current squeeze on capital financing and attendant high interest charges on money should not be unexpected. This being the case, it is not difficult to believe that financial investment problems could be alleviated appreciably following a cooling of the economy.

Probably the most imminent effect of a cyclic downswing is the impact of a smaller growth on the nuclear industry. With a net addition rate significantly below 10 GWe/yr for all generating capacity, a nuclear market of less than 5 GWe/yr might be expected in the U.S. prior to the year 2000. Much of this growth depression, however, has already been discounted since the number of orders for new plants has reached a dismally low level during the last few years. If growth acceleration is to begin around the year 2000, increasing market commitments should be expected around 1985, since some 15 years are required from commitment to plant commercial operations. The exact timing of this upswing may be uncertain, but if energy growth follows the pattern of past cyclic behavior, some upswing can almost certainly be expected within a five or ten year period on either side of the year 2000.

The growth problems for the electrical equipment supply industry would not appear to be so grim for the next 20 years in the rest of the world. Because of the position on the growth curve and because of a greater concern for energy independence, generating plant additions and particularly nuclear plant additions for the rest of the world will probably be an order of magnitude higher than those in the U.S. Though a continuing growth in the rest of the world may sound refreshing relative to the U.S. projections, that modest growth could lead to a growth crunch sometime shortly after the turn of the century--with commitment implications beginning to be felt in the 1990s. A net addition rate of over 600 GWe/yr around 2020, for example, could have profound implications on the equipment supply industry, financial institutions and the fuel supply industry. If even 50% of the world's generating capacity depended on coal fuel at that time, coal mining activities throughout the world would have to be at a level some 20 times that in the U.S. today. This growth in coal utilization could be even further exacerbated if coal were to become a source of liquid fuels for use by the transportation sector.

Before leaving the subject of electricity growth, it will be useful to translate this growth data into estimated installed capacities for nuclear plants. Except for nuclear plants already under construction, it will be assumed here that 50% of electricity capacity will be furnished by nuclear stations up to 2025. This estimate might be high for the U.S., because of its large coal resources, but might be low for the rest of the world. A cycle-adjusted-logistic growth is again assumed.

Projected U.S. and world nuclear capacities for the years 2000 and 2025 are summarized in table 3.1. For purposes of comparison, data are also shown from other sources. Since estimated nuclear capacities have been decreasing significantly in the last few years, only recent data sources have been selected for the reference points.

Table 3.1 Projections of U.S. and world installed nuclear capacities (GWe)

	U.S.		WORLD	
	CAL	DOE	CAL	TYPICAL
2000	176	325	620	1500
2025	1027	615	5820	6000

The referenced forecasts for the U.S. installed nuclear capacities are the median values of high and low DOE projections [6] made in 1978. A subsequent 1979 DOE forecast is actually available, indicating a newer projection of 235 to 300 GWe--a range still considerably higher than the CAL projection. In contrast, though, even the 1978 high estimate (910 GWe) for installed nuclear capacity in 2025 is lower than the CAL projection. Perhaps the relatively low DOE estimates for 2025 illustrates the tendency to project low installed capacities for the longer-range future based on shorter-range low-growth trends.

Projections for world nuclear capacity are more difficult to select uniquely. Some typical estimates for 2000 and 2025 are:

	2000	2025
INFCE [7] median for WOCA*	1000	2900
Häefele [8]	1500	8000
Giraud [9]	2000	5000
Kiely [10]	1540	5000 (in 2020)

*INFCE: International Nuclear Fuel Cycle Evaluation
 WOCA: World Outside Communist Areas.

Probably the values of 1500 and 6000 GWe for the years 2000 and 2025, respectively, typify these estimates. Again, the CAL projection for world installed nuclear capacity in 2000 appears to be much lower than traditional estimates. However, the CAL projection for 2025 appears to fall somewhat in the middle of the referenced estimates.

In summary, then, the growth of electricity consumption already is decreasing relative to the 1950-1975 period and, according to the CAL projection, can be expected to decrease still more. The decreasing number of electricity-generating stations being planned practically guarantees that electricity growth will continue to be small for another 10 to 15 years. Decreased nuclear growth is, of course, a product of that decline. But, a danger now exists that policy planning will reflect a presumption of a continuing low energy growth. Such a policy could be even more disastrous than the previous one of overestimating growth.

THE RESOURCE-DEMAND ISSUE

In 1974, the National Uranium Resource Evaluation (NURE) was established to review and assess uranium resources in the U.S. In addition to exploratory surveys conducted throughout the U.S., the NURE program has been augmented with confidential information from the uranium supply industry. On the basis of this program, the U.S. has probably done a more complete job of assessing uranium resources than any other country. While the NURE resource estimates have been criticized from time to time both for being too optimistic or too pessimistic, those resource estimates are probably the most reliable ones available. In April of 1979, the DOE estimates of uranium reserves and resources were as shown in table 3.2.[3] The estimates are classified both by degree of assurance and by forward costs. Uranium *prices* generally reflect not only the forward cost, but also sunk costs, the return on investment and the supply vs demand in the marketplace. Typically, the price of U_3O_8 is around twice the forward cost. Hence, a forward cost of $50 per lb U_3O_8 might suggest a price of about $100 per lb.

Similar estimates for the world have been prepared by the OCED and IAEA.* The world estimates tend to appear more pessimistic for several reasons, viz:
- they do not include categories corresponding to possible and speculative resources,

*OCED: Organization for Economic Cooperation and Development
 IAEA: International Atomic Energy Agency.

Table 3.2   U.S. uranium resources

Cost Category $/lb U$_3O_8$	Resources, 10^6 ST U$_3$O$_8$			
	Reserves	Probable Resources	Possible Resources	Speculative Resources
$\leq$ 30	0.690	1.005	0.675	0.300
$\leq$ 50	0.920	1.505	1.170	0.550
TOTAL				4.145

- they do not include resources in communist countries, and
- they are based on less thorough exploration than is the case in the U.S.

Table 3.3 summarizes these resource estimates for the non-communist world.

The Nuclear Energy Agency/International Atomic Energy Agency has established an International Uranium Resource Evaluation (IURE) project to assess world uranium resources. Based on their evaluations, they estimate a possible and speculative potential resource base of 8.5 to 19.2 million ST U$_3$O$_8$ in countries outside communist areas. Adding this to the reserves and possible resources, a total potential resource base of 14.2 to 24.9 million ST U$_3$O$_8$ is implied for the world outside communist areas. Adding their estimate of 4.3 to 9.5 million ST U$_3$O$_8$ in communist areas, a total world base of 18.5 to 34.4 million ST U$_3$O$_8$ is inferred.

Table 3.3   World uranium resources* (OCED/IAEA data for
            non-communist world)

Cost Category $/lb U_3O_8	Resources, 10^6 ST U_3O_8	
	Reasonably Assured	Estimated Additional
≤ 30	2.2	2.0
≤ 50	2.9	2.8
TOTAL		5.7

*Total resource base estimated by IURE is substantially
higher--see text.

   Obviously those estimates are very speculative and, at
best, only indicate the potential resources that *could* exist if
already discovered. Those estimates, however, are probably not
inconsistent with the U.S. data where much more extensive
exploration has occurred. If one assumes that resources in the
entire U.S. should be a reasonable sampling base for the world,
and that the U.S. land area is about 7% of the world land area,
one might expect a world resource base in the neighborhood of
50 million ST U_3O_8. Hence, the range of 18.5 to 34.4 million
ST U_3O_8 does not appear to be an unreasonable estimate.
   It has been a popular exercise in past years to compare
domestic uranium requirements to projected uranium resources
in future years to show that economically recoverable uranium
resources might be exhausted in only a few decades without
the prompt introduction of fast breeder reactors. This argument,
however, has lost much of its strength primarily as the result
of lower growth projections. Even with the revived nuclear
growth projected by the cycle-adjusted-logistic growth curve for

U.S. electricity growth, it is difficult to make a strong case for breeder reactors in the next 50 years simply to avoid the threat of uranium resource exhaustion. To illustrate this, cumulative uranium requirements for the U.S. are shown in figure 3.4 for both the once-through fuel cycle and recycle using only LWR plants during the next 50 years. Even for the once-through fuel cycle, resource exhaustion does not appear to be a problem in that time frame. The cumulative requirements shown in the figure have been based on the following assumptions:

- nuclear growth from now to the year 2000 is based only on present commitments (roughly consistent with CAL growth projections);
- beginning in the year 2000, half the net additions will be nuclear plants;
- uranium enrichment technology will be improved somewhat so that the residual uranium tails contain only 0.1% instead of 0.2% U-235; and
- modest improvements will occur in the efficiency of using uranium in LWR and other reactor plants.

Clearly, a larger fraction of nuclear plants could increase the cumulative requirements, though it appears that resource exhaustion would still not be a problem in the 50-year period. Beyond the 50-year period, the availability of high-grade uranium ore could become a problem, but speculations on energy technologies beyond 50 years are particularly inappropriate. This can be emphasized by noting that energy technology projections 50 years ago would have completely overlooked nuclear energy, since nuclear fission was not discovered until 1939 (some 40 years ago).

Figure 3.5 shows the similar situation for cumulative uranium requirements throughout the world. Again, it is assumed that 50% of the electrical generating plant additions are nuclear plants; an estimated fraction that could be too

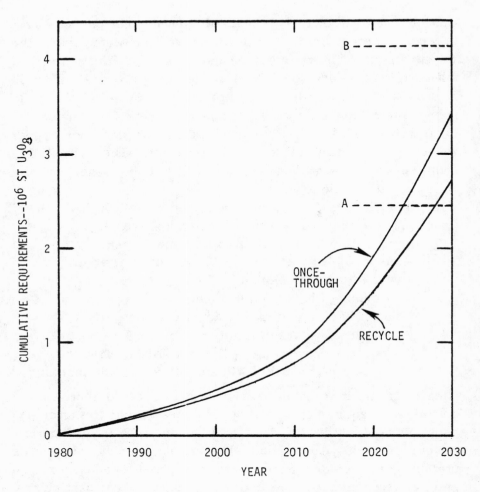

A: Reserves plus probable resources
B: Possible plus speculative resources

Figure 3.4   Projection of cumulative U_3O_8 requirements in the
U.S. assuming a CAL growth for nuclear energy

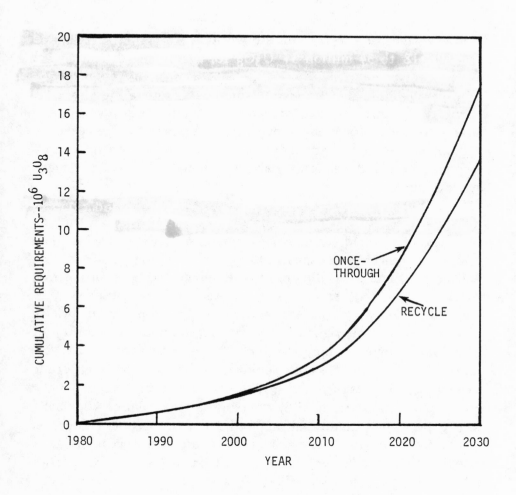

Figure 3.5 Projection of cumulative U_3O_8 requirements in the world assuming a CAL growth for nuclear energy

conservative for the rest of the world. Assuming a resource base of 18 to 34 million ST U_3O_8 for the world, the argument of resource exhaustion becomes a possible one, though apparently not a highly probable one. It is difficult, then, to justify the introduction of more efficient fuel cycles and alternative reactor designs even for the world by the resource exhaustion argument, though other arguments, such as energy dependence and supply capability may be much more compelling.

A more convincing argument for early introduction of more efficient fuel-cycle and reactor technologies, both in the U.S. and the world, comes from an examination of the annual supply requirements imposed on the mining and milling industry. Figure 3.6 illustrates the annual U_3O_8 requirements, first for the U.S., assuming a cycle-adjusted-logistic growth of nuclear energy consistent with the previous assumptions. Shown also in the figure is the projected supply capability of the U.S. uranium supply industry. It is estimated by DOE[7] that the maximum possible supply of uranium from high-grade resources will reach about 85,000 ST U_3O_8 per year after the year 2000. In addition, they conclude a maximum of 40,000 ST U_3O_8 per year could be supplied from low-grade resources, primarily from the by-product of phosphate fertilizers. It should be emphasized, however, that these are *estimates* of supplies that could possibly be produced.

The resulting requirements for the LWR once-through fuel cycle are somewhat higher than those developed by DOE, primarily because the data shown here assume the CAL growth that leads to an installed nuclear capacity of 1030 GWe in 2025 compared to the DOE assumption of 615 GWe at that time. The data shown here, however, assume a uranium tails assay of 0.10% for enrichment plants in 2025 compared to the very conservative DOE estimate of 0.20%.

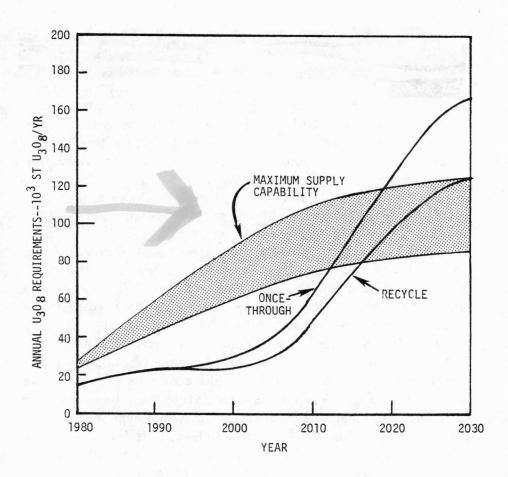

Figure 3.6   Projection of annual U_3O_8 requirements in the U.S.
assuming a CAL growth for nuclear energy

The data indicate that the requirements for a once-through cycle would exceed the maximum industry supply capability (including supply from low-grade resources) by the year 2020. Even with full recycle, the maximum supply capabilities would be challenged some ten years later. One can argue that the CAL growth assumption exaggerates the uranium needs. But, the main purpose for using this growth assumption is to illustrate how excessively conservative growth projections, based on today's growth trends, could lead to poor long-range planning strategies. And, indeed, that danger can increase still more if the growth over the period 1980 to 2000 follows the assumed CAL projection.

The situation is more worrisome for the world. The annual U_3O_8 requirements are again shown for the LWR once-through and full-recycle cases in figure 3.7. The supply-capability curves are based on DOE data[7] for the world outside communist areas, but have been increased by 35% to account for supplies inside communist areas. That ratio reflects the IURE estimate of total uranium resources for the world relative to the world outside communist areas. The data are obviously even more speculative than those for the U.S., but again indicate the world supply problem will require more attention than the U.S. one. Indeed, it is likely that the world situation will be even more serious than shown since it may be necessary for other parts of the world to supply more than 50% of their needs from nuclear energy.

In summary, then, the problem of resource exhaustion does not appear to be as serious as one of simply assuring an adequate annual supply of U_3O_8 to the utility industry. On the basis of the growth assumptions used here, more efficient nuclear technology should be made available in the U.S. sometime following the year 2010 and in the rest of the world, perhaps even before 2010. This suggests that technology development should be planned to allow for significant commercial introduction of improved technologies beginning shortly after the turn of the century.

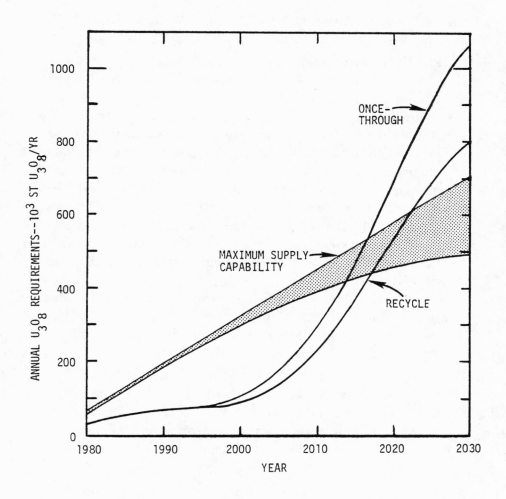

Figure 3.7   Projection of annual U_3O_8 requirements in the world
assuming a CAL growth for nuclear energy

OIL–RESOURCE AVAILABILITY AND NUCLEAR SUBSTITUTABILITY

Still once again, it is emphasized that the world energy crisis is, at this time, basically an oil supply/demand crisis. Moreover, for many industrial or neo-industrial countries having no significant oil, natural gas or coal resources, the problem is a fossil-fuel crisis. In the long range, of course, the fossil-fuel supply problem also will extend to all nations.   Hence, it is essential that careful attention be given to the development of technology for replacing our dependence on all exhaustible energy resources by inexhaustible energy resources.   While controversies continue on the necessity or lack of necessity for nuclear fuel recycle and breeding at an early date, the more urgent problem of finding ways to substitute nuclear technologies for oil-based technologies is getting insufficient attention at best.

Attention is, indeed, being given to a variety of synfuel-production technologies themselves based primarily on the use of coal resources. Although that effort is obviously important, it has at least two serious shortcomings, viz:

1. It does not offer a solution for most of the world where fossil fuels of all kinds are largely available *only as imports;* and

2. It is only a temporary solution since the supply of other fossil fuels will ultimately be severely pressed.

While the subject of transitions to inexhaustible energy resources is not explicitly a predominant theme of this book, it is, nevertheless, implicitly included--largely because of the projected strong growth of electricity and its probable dependence on nuclear resources.

Electricity substitution for space heating can be accomplished by the use of heat pumps, though our transfer from the use of fossil-fuel heat to electricity-operated heat (or solar-generated heat) may be relatively slow due to capital cost investments and the turnover rate of the housing stock.

Heat pumps and direct resistance heating using electricity may also be helpful in some industrial process-heat applications. However, the supply of economic high-temperature process heat will probably ultimately require the *direct* use of heat resources. Clearly, a market should evolve for nuclear-reactor process-heat plants to meet this demand--presuming technology is developed for this important application.

As has been previously emphasized, though, transportation and petrochemical feedstocks account for about 65% of the oil consumption. To make a significant impact on oil-supply requirements, then, substitutions must be found for these uses, particularly for transportation. At least three routes are available that allow the substitution of nuclear energy for oil, viz:

- nuclear-fueled process-heat reactors can be used to make synthetic fluid fuels (either liquid or gaseous);
- nuclear electricity-generating plants can be used to make hydrogen from the electrolytic decomposition of water; and/or
- stored electricity can be used for the propulsion of electric vehicles.

The first and last approaches would appear to be most economically attractive in the next few decades.

The importance of finding substitute portable fuels based on inexhaustible energy resources was emphasized in an ERDA (Energy Research and Development Agency) report published in September 1977.[11] Three particularly significant conclusions of that report were:

1. Presently planned inexhaustible energy systems are appropriate primarily for *electricity* production, but are not helpful to the problem of replacing oil as the primary source of portable energy.

2. A serious portable-fuel resource supply/demand gap will exist in the U.S. for the next few decades even with intensive conservation, accelerated coal production and maximum deployment of inexhaustible energy sources.

3. The inability of currently-known inexhaustible technologies to solve the portable fuels problem points up the critical need for re-orienting government research and development to development programs for an "economic portable energy system".

Because of the implications on international stability, one would think these recommendations would be accorded a policy attention at least as prominent as those being accorded the more passive policies of nuclear nonproliferation. Regrettably, that has not been the case so far.

It must also be recognized that greater reliance on nuclear fuel resources as an inexhaustible energy resource implies that nuclear fuels in the long range must depend on:

- breeder reactors,
- fusion breeder reactors, or
- more economic recovery of low-grade uranium resources.

At this time, only the availability of breeder reactors appears to be an assured source of inexhaustible nuclear fuels. If nuclear energy is to allow some relief from the current plight of oil supply/demand imbalances, the longer-range supply/demand problems associated with nuclear fuel resources cannot be ignored.

THE ECONOMIC ISSUE

Nuclear economics and economic-related technology issues will be a major topic of discussion in the following two chapters. But a discussion of changing issues would not be complete without some perspectives on the changing economics of alternative nuclear and non-nuclear energy-generating systems, and the implications of the changing economics on policy directions.

As was implied in the discussion of the uranium-resource demand issue, even that issue is basically an economic one. "Resource exhaustion" of elemental minerals is only a characterization of "resource degradation". In the discussion of cumulative uranium-resource requirements, for example, it was somewhat arbitrarily assumed that the consumption of all uranium with a forward cost of $\leq$ \$50 per lb U_3O_8 (a price of about \$100 per lb) constituted resource exhaustion. The more important point is that energy systems and systems policies should be selected to allow an optimum economic balance between *all* cost factors, with the requirement that the overall cost of a system be economically attractive relative to competitive systems (particularly with social costs included).

In general, the economics of fossil-fueled generating stations tend to be dominated by fuel-resource costs while the economics of nuclear plants tend to be dominated by fixed costs related to the plant capital cost. This is illustrated by figure 3.8, where the component costs for coal plants are compared to those for nuclear plants. Typically the cost contribution of coal fuel to the economics of coal-fired plants might fall roughly between 40 and 50% (or more) depending on the source of the coal and the transportation costs. The fuel-related cost contribution of oil and natural gas is generally even a larger fraction. In contrast, the fuel-cycle cost of a typical LWR plant contributes only about 30% to its generating cost; and the U_3O_8 resource expense contributes only about 15%, even with a once-through (non-recycle) utilization of the fuel. If the residual fuel (including plutonium) is recycled in the LWR, the resource-expense fraction is reduced to about 11%, but the added expenses of reprocessing and refabricating the spent fuel roughly

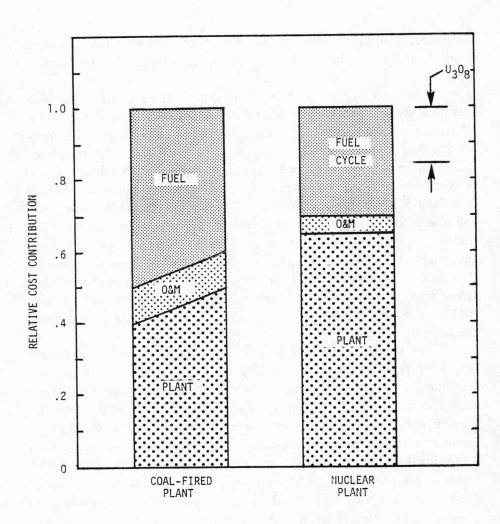

Figure 3.8   Relative contributions from fuel costs in coal–fired
            and nuclear plants

compensate for the smaller resource (and enrichment) expense, for U_3O_8 at about \$40 per lb.

With a more resource-efficient thermal-spectrum reactor, such as an advanced converter reactor, the expense associated with U_3O_8 consumption could be decreased to, perhaps, 6% of the generating cost. And, of course, the U_3O_8 cost contribution would decrease essentially to zero for a breeder reactor. The important conclusion is that a doubling of resource costs for a coal-fired station could increase its generating cost by 50%, while a doubling of U_3O_8 resource costs would increase the generating cost of a nuclear station 6 to 15% for thermal-spectrum reactors, and would have essentially no effect on the generating cost of a fast breeder reactor.

With fixed charges associated with plant capital costs contributing some 65% to the generating cost of contemporary nuclear plants and with uranium resources contributing only 10 to 15%, a significant sacrifice in capital cost cannot be justified under current conditions simply to improve the fuel resource expense. Indeed, a 50% increase in the capital cost would require a zero cost for all other operating costs to retain competitiveness. This, of course, is impossible, even with a breeding fuel cycle. However, if the U_3O_8 price triples or quadruples, for example, then it becomes easier to justify some capital cost penalty for a more resource-efficient reactor to reduce the U_3O_8 cost contribution.

If higher U_3O_8 prices are imminent, then the development of resource-efficient reactors, including breeder reactors is critical. If U_3O_8 prices are not expected to increase substantially in the near term, then primary attention should be directed toward improvements in the capital cost of nuclear plants. Hence, nuclear development policies depend strongly on potential price changes in U_3O_8.

   The Department of Energy policy planners have traditionally assumed that U_3O_8 prices will steadily increase as the
higher-grade uranium ore deposits are depleted and more
expensive lower-grade deposits must be mined. On this basis,
U_3O_8 prices are assumed to increase monotonically with
cumulative ore consumption. While this assumption might seem
logical, commodity prices generally do not behave in this
manner. Indeed, with other minerals it has typically been
found that improvements in recovery technology tend to offset
much of the added effort generally needed to recover the
lower-grade ores.
   More realistically, commodity prices usually respond
more to supply/demand balances (or imbalances) than to
cumulative resource consumption. With major shifts away from
oil and natural gas to coal and nuclear energy, it is likely
that growth demands for coal and U_3O_8 could put severe
pressures on both of these supply industries. The needs for
heavy capital investments, substantial expansions in industry
personnel, expansions in transportation equipment (especially
for coal) and more attention on the environmental implications
will all contribute to higher resource prices in the face of
rapidly increasing demands.
   Referring again to figure 3.6 in the previous section, it
would appear, then, that the excess supply capability over
demand for the next 20 years will probably result in some
softening of U_3O_8 prices. In fact, if spot market prices of
U_3O_8 are plotted in constant dollars (1979 dollars, for example) over the last two years, it can be seen that U_3O_8
prices are already falling somewhat. Qualitatively, it would
seem reasonable to expect some continuing softening of U_3O_8
prices, perhaps to $35 or $30 per lb (in 1979 dollars) in the
next ten to fifteen years. But, as a significant number of new
nuclear commitments becomes apparent around 2000, another
price escalation could occur, similar to the one that resulted
in the 1973-1975 period. Based on previous history, it would

not be surprising to see the price quadruple once more--resulting in a price of, perhaps, $120 per lb around 2010 or 2020. Assuming that the cumulative U_3O_8 requirements should reach about 2.0 million ST U_3O_8 by 2025 (as indicated by the CAL growth curve and LWR fuel recycle), the DOE price projection would suggest a mid-range price of $100 per lb U_3O_8. Hence, the assumption of a price of $120 per lb around 2020 appears to be rather consistent with DOE price projections, though the price trend projected for intermediate years would be quite different. The advantage of recognizing the supply/demand effect is that policy-makers might be less apt to underestimate prices some 30 to 40 years from now on the basis of a complacency that could develop during a weak market period for the next 10 to 15 years.

The U_3O_8 price projections suggest, then, that policy planning should be based on the assumption that more resource-efficient or more resource-insensitive reactors and fuel cycle technologies should be planned for significant development beginning about 2010. On a worldwide basis, the timing might actually be somewhat sooner. Just what those technologies might be will be the subject of Chapters IV and V.

Even though continuing attention on the improvement of reactor capital costs and operational utilization efficiency (or load factor) are important in the next 10 to 20 years, the requirement for the substantial introduction of resource-efficient technologies in some 30 years from now implies, then, that development and demonstration of these technologies must proceed at this time.

Before leaving the subject of economics, though, it is also useful to look at the changing issue of other costs associated with the nuclear fuel cycle. In particular, the changing handling or service costs associated with the fabrication of fresh fuel elements, the refabrication of recycle fuel elements and the reprocessing costs of spent reactor fuels will be examined.

Profound changes have occurred in the projected costs for some of the fuel-handling charges, just within the last seven to eight years. These changes are illustrated by figure 3.9 where fabrication, refabrication and reprocessing charges for the LWR are shown by projections made in 1971,[12] 1976 [13] and 1978. [14] Also shown is a weighted-average handling cost assuming 25% of the LWR fuel contains Pu/U mixed oxides. The overall handling cost shown in the figure also includes estimates of shipping and waste-management expenses.

While the fresh-fuel fabrication cost has increased at a rate approximately equivalent to that of normal inflation, the costs for refabrication and reprocessing have increased four to eight times. The overall LWR handling-cost projection for self-generated recycle has increased almost four times in seven years. Obviously, these cost increases do not enhance the economic incentive for fuel recycle in the LWR. Nor has this rapid increase been helpful to the projected economics of breeder and near-breeder reactors. However, a more quantitative evaluation of the incentive for recycle under various conditions requires a closer examination of all aspects of the fuel cycle economics. This will be done in the following chapters.

THE SAFETY AND ENVIRONMENTAL ISSUE

The potential environmental impacts associated with alternative energy technologies are more difficult to quantify than are resource-supply adequacies or energy economics. Yet the very survival of nuclear-energy technology depends on the public's perception of nuclear safety and environmental risks. The two overriding concerns are those related to:

- reactor operating safety, and
- nuclear waste management.

In both cases, the isolation of fission products from the surrounding environment is essential.

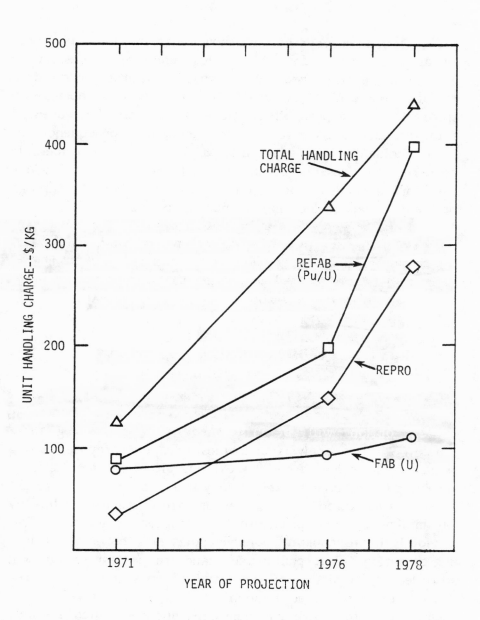

Figure 3.9   History of projections for handling charges (LWR)

Opinion surveys [15] have shown that approximately two-thirds of the public feel that nuclear energy is necessary to meet future energy needs, but another two-thirds would object to siting nuclear plants within five miles of their homes. In spite of the low probabilities generally claimed for potential fatalities arising from reactor accidents, the consequences of such accidents are apparently a subject of considerable concern. Perhaps one of the most effective ways to reduce the consequences of potential nuclear accidents is through policy decisions on locations of nuclear plants.

In an analysis by the Institute for Energy Analysis, [16] a nuclear siting policy is suggested that would lead to a relatively few number of nuclear generating centers. Such a policy would have several advantages including the following:

- land areas vulnerable to contamination could be minimized;
- security could be simplified;
- transport of fuel could be minimized; and
- competent staffs could be aggregated.

By locating these sites away from populated centers, much of the public concerns could be eliminated and, indeed, the nuclear option will be in jeopardy without allaying those concerns. The analysis concludes that the proposed siting policy could evolve from the existing situation since a large number of the current sites already meet appropriate criteria.

As indicated earlier, reactor safety has leaned heavily on minimizing the probability of reactor accidents. This obviously is an important objective that must receive continuing attention. But, public acceptance of nuclear energy will also depend on minimizing the potential consequences of reactor accidents. A more specific siting policy is probably the easiest way to assure minimum consequences. Such a policy would, of course, also have important implications on institutional aspects of the energy industry--some of which will be discussed in the last chapter. But, it could contribute to the survival of the nuclear option and to energy survival itself.

The other significant problem related to the environmental issue is that of nuclear waste management. In particular, the significant disagreements and policy indecisions regarding the ultimate disposal of nuclear-fuel waste products have been very damaging to public acceptance. The problem has become even more serious in the last few years because of some state mandates in the U.S. and some legislative restrictions in several other countries that have established a moratorium on further nuclear growth until satisfactory waste-disposal facilities are assured.

Both fossil-fueled plants and nuclear plants must be concerned about waste disposal. On the one hand, wastes from fossil-fueled plants are large and are handled primarily by release to the biosphere with dependence on atmospheric dilution and dissemination. On the other hand, wastes from nuclear plants are small, but more toxic on an equal-weight basis. While this higher specific toxicity creates important handling problems, at least the wastes from nuclear plants can be contained. In fact, if nuclear wastes can be stored with an average density of about one gram per cubic centimeter, the volume of wastes from a large nuclear plant operating for one year could be contained in a package one meter on each side. Even at somewhat lower waste densities, to allow for heat removal, containment is clearly feasible.

To gain public acceptance for the use of nuclear energy, though, a clear, agreed-upon plan for waste-disposal demonstration is required. Without that plan, one important option for energy survival may be lost.

THE INSTITUTIONAL ISSUE

Perhaps one of the most profound lessons to be learned from the history of the nuclear industry is that the technology development costs, the front-end commercialization costs, the

economic risks and the political risks associated with the
business are simply beyond the capability of private enterprise
as currently structured. Yet the importance of a continuing
strong national energy supply is obvious. In the past, the
institutional problems have been solved by a hodge-podge of
expedients, including:

- spin-off technology from military programs
  (e.g., the nuclear navy),
- AEC/ERDA/DOE R&D support to national
  laboratories and industries,
- government/utility/supply-industry demonstration programs, etc.

In spite of these industry-assistance expedients, it is unlikely
that a major new technology, such as the LMFBR, could be
brought through all the necessary stages of commercialization
without public outcries of government bail-out or industry
outcries of imprudent risk assumptions.

It is tempting to believe that a new cyclic surge of
technology growth could lean as much on innovative
institutional approaches to the societal energy problems as on
innovative technology itself. It would seem that new solutions
to technology development, demonstration and deployment all
must be found. While the introduction of new reactor
technologies tends to dominate our thinking, it is likely that
the continuing development and commercialization of the *fuel
cycle* may be an even more formidable problem. The difficulty
of establishing a fuel-cycle "infra-structure" has frequently
(and probably justifiably) been singled out as a rationale for
pursuing only the already-initiated plutonium recycle industry.
Yet this rationale is basically a confession of institutional
impotence.

In addition to the technology development and implementation problems, new issues involving financing, security,
safety and non-proliferation concerns must be resolved. Par-

ticularly if international or multinational efforts are involved, the institutional problems can be complicated still more.

More on this subject will be included in the last part of this book. But, the problems are formidable and clear solutions are probably impossible at this time. This only emphasizes the urgency of attacking the institutional problems as promptly and conscientiously as possible.

REFERENCES

1.   "Nuclear Power Issues and Choices", Sponsored by the Ford Foundation, Administered by the MITRE Corporation, Ballinger Publishing Company, Cambridge, Mass., 1977.

2.   "Cost-Benefit Analysis of the U.S. Breeder Reactor Program", U.S. Atomic Energy Commission, WASH-1184, January 1972.

3.   Nininger, Robert D., "Uranium Resources and Supply", presented at Atomic Industrial Forum, New York, March 1978.

4.   "Statistical Data of the Uranium Industry", U.S. Department of Energy, GJO-100(79), January 1979.

5.   "Uranium Supply Outlook Assessment", Study Conducted by S. M. Stoller Corporation for the U.S. Department of Energy, February 1979.

6.   "Annual Report to Congress 1978", Energy Information Administration, DOE/EIA-0173/3.

7.   "Nuclear Proliferation and Civilian Nuclear Power", U.S. Department of Energy, DOE/NE-0001, June 1980.

8.   Häefele, Wolf, "Global Perspectives and Options for Long-Range Energy Strategies", Energy, Volume 4, pp. 745-760, January 1979.

9.   Giraud, Andre, "World Energy Resources", presented at Conference on World Nuclear Power, Washington, D.C., November 1976.

10. Kiely, John R., "Energy-Looking Into the 21st Century", presented to the American Power Conference IEEE Luncheon, Chicago, Illinois, April 1979.

11. "The Need for and Deployment of Inexhaustible Energy Resource Technologies", Report of Technology Study Panel, U.S. Energy Research and Development Administration, September 1977.

12. "Reactor Fuel Cycle Costs for Nuclear Power Evaluation", Prepared for Division of Reactor Development and Technology, U.S. Atomic Energy Commission, WASH-1099, December 1971.

13. "Final Generic Environmental Statement on the Use of Recycle Plutonium in Mixed Oxide Fuel in Light Water Cooled Reactors", U.S. Nuclear Regulatory Commission, NUREG-0002, August 1976.

14. "Fuel Cycle Cost Studies-Fabrication, Reprocessing, and Refabrication of LWR, SSCR, HWR, LMFBR, and HTGR Fuels", Oak Ridge National Laboratory, ORNL/TM-6522, March 1979.

15. Pokorny, Gene, "Living Dangerously--Sometimes", Published in Public Opinion, June/July 1979.

16. "Economic and Environmental Impacts of a U.S. Nuclear Moratorium, 1985-2010", Institute for Energy Analysis, MIT Press, 1979.

Part Two

TECHNOLOGY DIRECTIONS

Chapter IV

Nuclear Technology Evolution:
The Prevalent Strategy

To be reviewed in this chapter is the prevalent strategy that has been traditionally pursued for nuclear-technology evolution. In the succeeding chapter, a "not-so-prevalent strategy" will be introduced. One objective of the review in this chapter is to identify the *problems* associated with the prevailing evolutionary strategy--particularly the now-apparent economic disincentives that threaten to impede its implementation. The strategy to be described in the subsequent chapter will, then, suggest an alternative approach that could remove at least some of the economic disincentives.

As an introduction to the discussion, it will be useful, perhaps, to identify both those considerations that led to the prevalent strategy and those that did not. Primary considerations that shaped nuclear policies some ten years ago were:

1. a concern that nuclear energy growth would exceed the supply capability of the uranium mining and milling industry, and
2. a belief that the economics of breeder reactors (and associated fuel recycle) would justify their early development and deployment.

While both of those premises can no longer be accepted as valid for the next 20 to 30 years, it is likely that a resurgence of energy growth and its consequent heavy demands on uranium could, once again, restore the validity of those premises. And, indeed, such a change could occur quite abruptly. Because of the relatively long time required to develop and demonstrate appropriate new technologies and, particularly, because of the time to deploy a large enough base of new technologies to make a significant impact, a real danger now may be the complacence that could occur during the intervening period when technology development and demonstration should be proceeding.

But, it may be equally useful to consider the considerations that did *not* shape nuclear policies a decade ago, but which now appear to be crucially important. Those considerations are:

1. the public acceptance of nuclear energy as a serious problem was not adequately recognized;
2. a political concern that civilian nuclear energy could serve as a pathway for nuclear-weapons proliferation was clearly underestimated; and
3. any need for nuclear energy as a potential technology resource for solving a future portable-fuel problem was almost completely unrecognized.

Each of these considerations has subsequently become exceedingly important, although it is understandable they could have been overlooked as important concerns some ten years ago. Yet, in view of probable future U.S. and world energy problems, it is also quite clear that abandonment of nuclear energy is not an appropriate response to the public-acceptance, non-proliferation and portable-fuels concerns.

The growing acceptance problem associated with nuclear energy is not a unique one. Public acceptance of large industrial complexes, in general, has become a problem--particularly those industries that are viewed as potential sources of environmental degradation to local communities. Probably future societies will solve the more difficult industry acceptance problems partly through still more stringent requirements on their environmental impact, but largely by greater attention to land-use zoning, i.e., requiring more separation between industrial and residential communities.

There is also a growing realization that the nuclear-weapons-proliferation concern is very real and serious. But, the denial of civilian nuclear energy as a potential energy resource to the newly-developing industrial nations could be equally as threatening to world stability. Probably this problem will have to be solved primarily by institutional mechanisms.

It is now becoming apparent, as emphasized in the previous chapter, that the solution of the portable-fuel problem may be the most critically important of all the formerly unappreciated concerns. At this time, then, each of the unrecognized (or under-recognized) problems has become more critical than the recognized ones. Consequently, some attention must be given to these newer problems.

However, since this chapter deals primarily with the prevalent strategy, which had its genesis some ten years ago, the discussion here will center on the question of resources and economics. In the following section, the general features and economic objectives of the prevalent strategy will first be discussed. The succeeding sections will then review the projected cost performance of the light water reactor (LWR), the advanced converter reactor (ACR) and the fast breeder reactor (FBR), based on probable economic trends over the next few decades.

THE PREVALENT STRATEGY AND ECONOMIC BACKGROUND

The prevailing strategy governing nuclear-technology planning has traditionally been characterized by the following steps:

- use of the low-enrichment-uranium (LEU)
  once-through fuel cycle in LWR plants initially;
- use of self-generated plutonium (Pu) recycle
  when a recycle industry evolves;
- use of the thorium (Th) cycle with self-
  generated U-233 recycle in advanced
  converter reactors, when appropriate;

- re-assignment of LWR-generated Pu to FBR
  plants as the latter become commercialized;
  and
- self-sustaining operation of FBR plants on Pu
  fuel in the very long range, with surplus Pu or
  U-233 being supplied to satellite thermal-
  spectrum reactors.

It was generally presumed by policy planners that the evolution of this strategy would be expedited by the promise of improved economics, particularly in the face of higher U_3O_8 prices. Admittedly, it was anticipated some introduction expenses might be required during the early phases of the evolutionary steps, but it was expected that learning experience and a large volume of business would quickly compensate the front-end investments.

In view of the changing economic conditions surrounding nuclear energy, those assumptions must be re-examined. The economic performance of LWR plants using the LEU once-through cycle will form the basis for the economic assessment of technology alternatives, simply because the LWR technology has already been accepted as economically attractive by the utilities, at least under current conditions.

To assess the future economic incentive for alternative nuclear technologies, it is useful to identify some evaluation criteria. In particular, it will be useful, for perspective, to define a benchmark U_3O_8 price and a probable range of enrichment prices that might reasonably be expected in 30 to 40 years. Finally, a plausible generating cost limit, that might be regarded as acceptable should be identified. These selections are, of course, largely judgemental.

The price of U_3O_8 has increased by a factor of about six times in the last decade, or about three times when measured in constant dollars (e.g., 1979 dollars). While it is now clear that the exhaustion of U_3O_8 resources with forward costs up to $50 per lb is *not* credible within the next 30 years, it *is* credible that prices could again treble during another energy-growth surge. For this reason, a U_3O_8 price of $120 per lb (1979 dollars) will be assumed as a plausible benchmark price for the economic comparison of alternative systems.

Some attention should also be given to the possible impact of changes in separative work costs. With U_3O_8 at $40 per lb and enrichment at $100 per SWU (separative work unit), about half the cost of U-235 in low-enrichment uranium arises from the ore cost and half from the enrichment cost. For increases in U_3O_8 costs and decreases in separative-work costs, it becomes economically advantageous to apply more separative work in the enrichment process such that more of the U-235 is recovered. The optimum amount of separative work (for the optimum enrichment-tails assay) is a complicated function of the component costs. But, it is illustrative that a decrease of 50% in enrichment cost would offset about 30% of the U-235 cost rise resulting from a U_3O_8 price change from $40 to $120 per lb. Since such an enrichment price improvement might be expected from advanced isotope separation technologies by 2010, the effect of this probable cost change will also be examined in the subsequent economic evaluations. One consequence of this is that the commercialization of FBR plants will become more difficult. This is an important point frequently overlooked in long-range economic assessments.

With plausible price changes in U_3O_8 and separative work, then, it is necessary to examine how generating costs of alternative plants might be affected relative to an acceptable upper limit. If one assumes the price of coal might increase by some 50% during the next energy-consumption surge, a projection that appears to be conservative relative to recent history, then the generating cost of coal-fired plants would be expected to increase approximately 25%. On this basis, which is admittedly speculative, a prudent target for nuclear plants might be an upper-limit increase of 20% in generating cost, though some might argue that 10% could be a better target.

Once again, it is emphasized, then, that the purpose of selecting:
- a plausible high-demand U_3O_8 price,
- a plausible range of enrichment prices, and
- a maximum acceptable generating-cost penalty

is simply to provide a framework for economic examinations. As a basis for the subsequent comparison of the economics for alternative reactor systems, the base-case LWR economics will first be examined.

THE LWR ECONOMICS WITH ALTERNATIVE FUEL CYCLES

The economic performance analysis of electricity-generating plants is traditionally divided into a plant-related cost, an operating and maintenance (O&M) cost and a fuel or fuel-cycle cost. To assess the relative performances of alternative types of generating plants, it is convenient to measure the cost of electricity generated at the plant per unit of output, i.e., mills cost per kilowatt-hour of electricity output. Very approximately, the price of electricity to a customer is twice the generating cost because of transmission and administration costs. Hence, a penalty of 20% in generating costs, for example, would only represent a 10% increase in price to the customer.

The O&M and fuel costs are generally accumulated quite regularly throughout a year of operation and, therefore, can simply be divided by the annual output measured in killowatt-hours to arrive at an annual average generating cost for these cost components. The contribution from the plant cost is somewhat more difficult to identify. Interest charges on investment money and taxes are again relatively easy to measure in costs per year. But interest money is complicated by several factors, including the following:

- the expected rate of return on equity money is different from the interest rate on debt money;
- appropriate interest rates on debt and equity financing and on discount rates are all strongly affected by inflation rates of the general economy; and
- investment financing precedes by several years the revenue.

In the succeeding assessments, it will be assumed that the capital cost of an LWR plant is typically $800 per kwe (1979 dollars), that the annual charge rate against the final plant cost must be 10% per year,* and that the plant operates over its useful life 70% of the time (average load factor). The generating cost of the plant, assuming U_3O_8 at $40 per lb and enrichment at $100 per SWU would be typically as follows:

Plant:	13.0 m/kwh
O&M:	1.0
Fuel Cycle:	<u>6.0</u>
	20.0 m/kwh

The generating costs for alternative LWR fuel cycles will be compared, then, assuming that other fuel cycles affect only the fuel-cycle-cost component of the generating cost. (See Appendix B for more details on neutronics and fuel-cycle-cost discussions.) To simplify the comparisons, only the relative generating costs will be compared, assuming the generating cost for the base-case LWR with the once-through fuel cycle is 1.00.

*This annual-charge rate is typical in a zero-inflation economy.

The prevalent fuel-cycle strategy for the LWR has generally presumed the use of the LEU once-through fuel cycle in the near term with the introduction of self-generated recycle of plutonium when the economics of recycle could justify that step.   Some limited attention has also been directed in the last few years, to the use of the thorium cycle in the LWR.   The calculated relative generating costs for these three fuel cycles are illustrated in figure 4.1, where the U_3O_8 price has been assumed to be \$40 per lb and the enrichment price \$100 per SWU. In these examples, appropriate values have been assumed for fissile plutonium and U-233 relative to the value of U-235.   The "neutronic" value of plutonium is only slightly smaller than that of U-235, but its "market" value is degraded somewhat more by the higher cost of fabricating Pu- containing fuel.   While the neutronic value of U-233 is actually somewhat higher than that of U-235, its market value is penalized substantially, at least for use in LWR plants, by a very large cost for fabricating LWR fuel in shielded and automated process facilities.

There are two general observations worth noting from figure 4.1. These are:

1. There is essentially no economic incentive, under current economic conditions, to recycle residual fuel in the LEU fuel cycle; and

2. There is an economic disincentive of about 15% in fuel cycle cost and 5% in generating cost for the use of the highly-enriched uranium/ thorium (HEU/Th) fuel cycle in LWR plants, as presently designed, under current economic conditions.

In spite of the fact that the recycle depletion cost is reduced more than 30% over that for the once-through cycle, the increase in the other fuel-cycle component costs, (associated with plutonium/uranium refabrication, with spent-fuel repro-

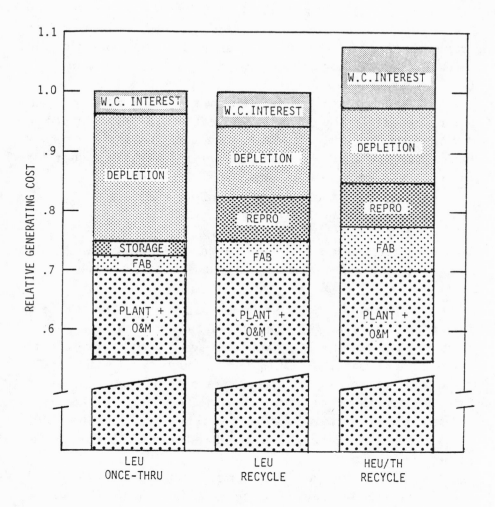

Figure 4.1 Relative generating costs for alternative LWR fuel
cycles (U_3O_8 at $40 per lb; enrichment at $100
per SWU)

cessing and with the additional interest charge against the bred fuel), compensate for the more economic utilization of uranium with fuel recycle. As will be seen subsequently, the incentive for plutonium recycle does, however, improve for U_3O_8 at higher prices.

The economics for the thorium fuel cycle reflects even more dramatically the penalties associated with higher refabrication costs and interest charges on the fuel. The refabrication cost, in particular, is significantly higher than that for the plutonium recycle because:

- more bred fuel must be recycled,
- some residual U-235 is mixed with the U-233, and
- the refabrication process is more difficult due to the presence of high-gamma-activity products in the U-233.

In figure 4.2, the relative generating costs for the LEU fuel cycles are again illustrated assuming that the U_3O_8 price rises to $120 per lb. The data have again been normalized to a value of unity for the LEU once-through cycle with a U_3O_8 price of $40 per lb, in order to show the generating-cost increase resulting from the higher U_3O_8 price. For example, it can be seen that the generating cost for the LEU once-through cycle has been penalized some 27% due to the higher U_3O_8 price.

For the higher U_3O_8 price, the fuel-cycle cost associated with full recycle of residual fuels in the LEU cycle clearly shows an improvement over the once-through fuel cycle. The fuel-cycle cost improvement for the recycle case is about 13% relative to the once-through cycle; and the generating-cost improvement is about 5%.

Such a large margin of improvement would certainly justify fuel recycle if that decision were based only on economic considerations. It is also noted that fuel recycle

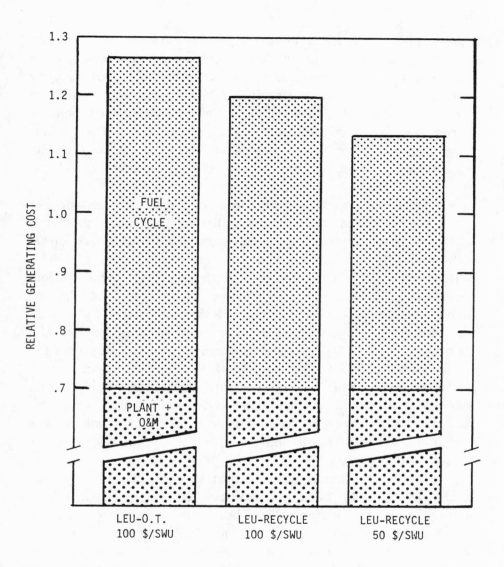

Figure 4.2   Relative generating costs for alternative LWR fuel
cycles and enrichment costs (U_3O_8 at $120 per lb)

would limit the generating-cost penalty associated with the higher U_3O_8 price to about 20%, which would apparently make this cycle just acceptable relative to the acceptability criterion that was arbitrarily defined earlier.   For an enrichment cost of $50 per SWU, the generating-cost penalty is reduced to about 13% as shown in the figure.   Though not shown, the thorium cycle becomes competitive with the LEU once-through cycle, but not with the LEU recycle, for U_3O_8 at $120 per lb.

More importantly, both the LEU and thorium self-generated recycle cases suffer from political concerns associated with the possible diversion of weapons-grade fuels. A 1000 MWe LWR plant would typically recycle some 250 kg of fissile plutonium each year. Likewise, the thorium cycle would require makeup fuel requirements of almost 400 kg of highly-enriched U-235 each year. If these sensitive fuels must be transported to some 100 or more reactor sites in the U.S., then both cycles, as presently conceived, would be regarded as unattractive. The sensitivity of the thorium cycle could be improved by the use of medium-enriched uranium (MEU) makeup fuel (with a maximum U-235 enrichment of 20%). However, this expedient would render the economics of the thorium cycle still more unattractive for use in the LWR as presently designed.

One must conclude, then, that the economics of *recycle* in the LWR could become attractive if the price of U_3O_8 should increase substantially, but the prevalent recycle strategies would still suffer some political concerns associated with national or sub-national diversion of sensitive fuel materials.   Probably some combination of institutional and technological approaches would be required to resolve that political concern.   More on that subject will be postponed until later chapters.

THE ADVANCED-CONVERTER-REACTOR ECONOMICS

The term "advanced converter reactor" (ACR) has been applied to thermal-spectrum reactors designed to be more resource-efficient than the LWR.   Included in this category are the:
- light water breeder reactor (LWBR),*
- high temperature gas-cooled reactor (HTGR), and
- heavy water reactor.

Usually it is assumed that the ACR uses the thorium fuel cycle, though that need not be the case. The preoccupation with improved uranium utilization is illustrated by the fact that ACR development has been almost completely directed toward fuel utilization in contrast to different end-use applications.

   In general, the thorium cycle appears to be more economically attractive in a well-designed ACR than was indicated for the direct replacement in LWR plants. In fact, the generating cost for a good ACR plant usually is projected to be competitive with the base LWR plant even for U_3O_8 at $40 per lb and a 5% penalty assigned to the capital cost of the ACR.   Although there tends to be some differences in economic performance for alternative ACR technologies, HTGR data will be used to illustrate the ACR economics in this examination.

   ACR generating costs are shown in figure 4.3 for U_3O_8 at $120 per lb.   The bar at the left side illustrates the LWR generating cost at the same U_3O_8 price assuming self-generated recycle of the bred fuel. In this illustration, the capital cost of the ACR has been assumed to be 5% higher

*The LWBR is generally classified as a breeder reactor. However, a more economically optimized version of this class of reactor would have a conversion ratio of about 0.9. It is this version that will be included here as an ACR.

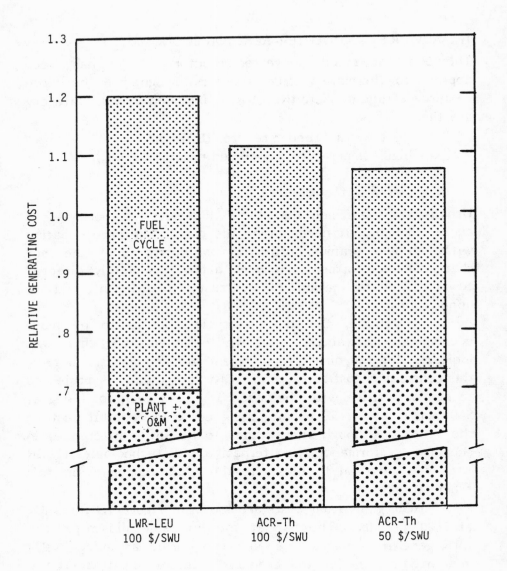

Figure 4.3   Relative generating costs for LWR and ACR plants
with fuel recycle (U_3O_8 at $ 120 per lb)

than that of the base LWR. The real economic merit of the ACR plants using the thorium cycle becomes apparent for U_3O_8 at elevated prices. But, as noted previously, a simple substitution of the thorium cycle in the current LWR plants does not appear to be attractive.

With U_3O_8 at \$120 per lb, the generating cost for the ACR is some 7% less than that of the LWR with recycle, and the overall penalty arising from the U_3O_8 price increase is only 13% even with the 5% capital-cost penalty. This is comfortably below the 20% criterion set for acceptability. As also shown in the figure, a simultaneous reduction in enrichment price from \$100 to \$50 per SWU would decrease the new generating cost to only 8% above the base case. It is very likely, indeed, that enrichment costs would be reduced by at least this amount due to advanced isotope separation technologies around 2010. This is an important observation because it is this system that may set the economic target for the fast breeder reactors.

For both ACR and FBR plants, savings in fuel costs frequently come at the expense of higher plant costs. Moreover, the competitive advantage of an advanced reactor system may only become assured at some elevated U_3O_8 price where the fuel-cycle cost advantage may be sufficient to offset the plant cost penalty. It is useful, then, to illustrate how this competitiveness changes with U_3O_8 price. One rather convenient method for doing this is to plot the ratio

$$\frac{(FCC)_x}{(PC)_{LWR}} = \frac{(\text{fuel cycle cost})_x}{(\text{plant cost})_{LWR}}$$

as a function of U_3O_8 price. The difference

$$\frac{(FCC)_{LWR}}{(PC)_{LWR}} - \frac{(FCC)_{x}}{(PC)_{LWR}} = \frac{\Delta}{(PC)_{LWR}}$$

represents the fractional plant cost increase (relative to the LWR plant cost) that could be tolerated to just offset the fuel-cycle-cost improvement shown by the alternative system x.

The results for the ACR (the HTGR in this case) are shown in figure 4.4 relative to the LWR, assuming self-generated recycle for both reactor systems. Hence, if the enrichment cost is, say, $50 per SWU for both cases, an ACR plant cost penalty of 6% could be afforded for U_3O_8 at $40 per lb, 10% for U_3O_8 at $80 per lb and about 15% at $120 per lb. Likewise, it can be seen from the figure that an enrichment cost improvement from $100 per SWU to $50 per SWU would allow the LWR to accommodate a price change in U_3O_8 from $40 to about $60 per lb with no economic penalty.

Once again, both the LWR and ACR self-generated recycle suffer from the political concerns associated with the use of sensitive fuel materials, i.e., U-235 and U-233 for the ACR and Pu for the LWR. It was previously noted that the LWR once-through fuel cycle is competitive with recycle for U_3O_8 at $40 per lb. Likewise, the MEU/Th once-through fuel cycle in the ACR appears to be approximately competitive with the HEU/Th self-generated recycle for U_3O_8 at $40 per lb. Consequently, ACR implementation plans could proceed with the less sensitive MEU initial fuel and no bred-fuel recycle until institutional arrangements for a recycle industry could be developed and deployed.

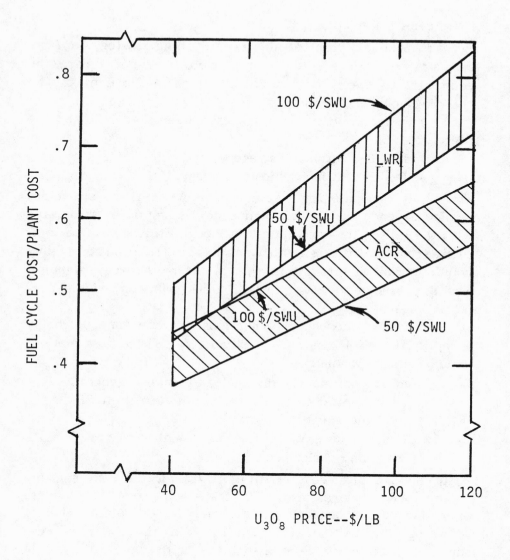

Figure 4.4   (Fuel cycle cost)/plant cost) ratio as affected by
U₃O₈ prices (self-generated recycle fuel management)

FBR ALTERNATIVE FUEL CYCLES

Perhaps most important of all, the prevailing strategy for fast-breeder reactors should be reviewed. In particular, the economics of the FBR must be examined in the light of the changing economic conditions.

Two traditional strategies have been suggested for the FBR, viz:

- the FBR autonomy strategy, and
- the FBR/ACR symbiosis strategy.

The usual rationale for a selection between these strategies would depend largely on economic assumptions associated with the FBR plant-capital cost, although other considerations, such as reactor safety, sitability and non-proliferation constraints could become significant factors in future strategy planning. The emphasis in this section will focus primarily on the economic imperatives.

Several economic assumptions, at one time, strongly favored the FBR autonomy strategy. Those assumptions included the following:

- a low capital cost for nuclear plants in general,
- a low capital cost difference between the LWR and FBR, and
- high LWR fuel costs associated with high-cost U_3O_8 and enrichment.

Assuming a capital-cost penalty of 10% for the FBR relative to the LWR, national cost/benefit studies,[1] some seven years ago, showed a complete domination of FBR plants shortly following the turn of the century. In these studies, the superior fuel-cycle cost associated with the FBR easily compensated for the relatively small capital-cost penalty assigned to the FBR.

When it subsequently became apparent that the capital cost of nuclear plants was rising more rapidly than inflation in general, and, more particularly, when it became apparent that the capital-cost penalty of the LMFBR was probably at least 25% above that of the LWR, the FBR/ACR symbiosis strategy became more popular. The symbiosis strategy viewed the FBR more as a fuel factory with advanced converter reactors as the work-horse power plants in the longer range. Presumably, a sufficient number of FBR plants would be utilized to assure a surfeit of bred-nuclear fuel. Under these circumstances, a glut of bred fuel would render the fuel cost of all reactors insignificant, therby making the capital cost the dominant economic consideration. The primary purpose, then, of the FBR would become one of assuring cheap nuclear fuel.

Under these circumstances, the most economic combination of reactors would be one where the number of cheaper ACR plants would be maximized relative to the FBR plants. Fortescue,[2] who has been a strong proponent of the symbiosis strategy, has shown that the ratio of ACR/FBR plants could be maximized by maximizing both the conversion ratio of the ACR and the breeding ratio of the FBR. With an FBR breeding ratio of 1.4 and an ACR conversion ratio of 0.9, for example, approximately four advanced converter reactors could be supported by each FBR plant.

To achieve a high ACR conversion ratio, it would be advantageous to use the thorium fuel cycle with U-233 makeup fuel supplied from the blankets of the FBR fuel factories. With an efficient advanced converter reactor, a conversion ratio of 0.9 with this fuel cycle should be easily achievable. While Fortescue has directed his attention toward gas-cooled reactors, Cohen[3] has made equally compelling arguments involving a symbiotic combination of LMFBR and advanced LWR plants.

The earlier autonomous strategy for the FBR presumed the use of plutonium fuel and U-238 fertile material in the core with blankets consisting entirely of natural or depleted uranium. The more recent symbiotic strategy would utilize essentially the same core composition but would use only enough uranium in the blankets to assure a self-sustaining plutonium supply for the FBR plants. Some of the blanket material, then, would consist of thorium to breed U-233 for use in the satellite ACR plants.

The emphasis in the study of symbiotic reactor systems has traditionally been on fuel resource balances between the FBR and ACR plants, though the motivation for the strategy was clearly an economic one. Primary emphasis in the discussion here will be directed toward the economic performance of the breeder reactor and, particularly, the economic incentive for introducing the FBR. Some overview of the FBR economic performance is illustrated by figure 4.5. Here the generating costs for the two fuel cycles are illustrated, again relative to a value of unity for an LWR plant, assuming U_3O_8 at $40 per lb. The capital cost for the FBR plant is assumed to be 50% above that of the LWR. This penalty is somewhat larger than that generally assumed for a well-established FBR supply industry, but is probably low relative to costs that might be expected during the early introduction period.

It can be seen from the figure that the fuel-handling cost is a much more important factor in the FBR fuel-cycle cost than is the fuel-supply cost component. The fuel-supply cost is simply the difference between the working-capital interest charges and the depletion credit. For a breeding ratio of 1.2, the production of fuel would be, typically, 3% per year of its total inventory. With a working-capital interest rate of 8%, the depletion credit would compensate about three-eights of the working-capital cost. It has been assumed in the figure that the value of plutonium is 50% that of highly-enriched U-235; and the value of U-233 is 120% that of U-235. Clearly, the revenue from the more valuable U-233 is quite beneficial for the FBR with a thorium blanket. Obviously, though, the production of some U-233 fuel presumes a market for the fuel.

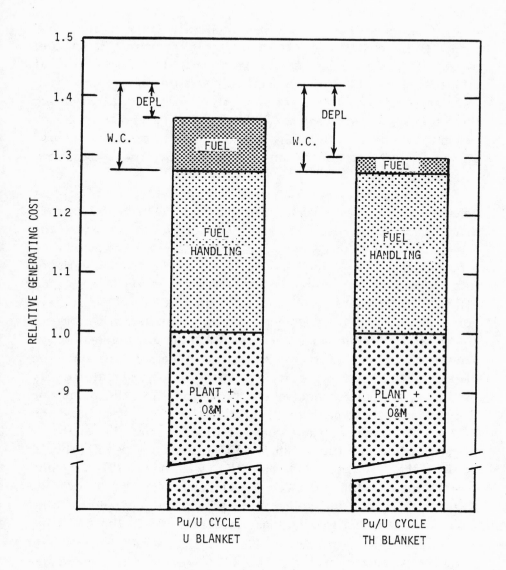

Figure 4.5   Relative generating costs for FBR with alternative
blanket choices (U_3O_8 at $ 120 per lb; enrichment
at $ 100 per SWU)

It is significant that the generating costs for both cases are uncompetitive, even with the LWR once-through fuel cycle. That conclusion is consistent with the observation of others that the LMFBR will probably not be competitive with LWR and ACR plants for the next 30 or 40 years; at least for the prevalent strategies. Indeed, the economic results presented here are generally consistent with those developed by the DOE Nonproliferation Alternative Systems Assessment Program (NASAP). [4]

As was previously noted, an arbitrary capital-cost penalty of 50% was assumed in figure 4.5. (The NASAP study assumed 35%.) A more useful approach is to examine the magnitude of a capital-cost penalty that could be allowed as a function of the U_3O_8 price. Figure 4.6 illustrates the ratio of fuel cycle cost to plant cost, as affected by U_3O_8 price, for the FBR relative to both the LWR and ACR. Results have been shown for two enrichment prices, consistent with current prices and possible future prices assuming advanced-isotope-separation technologies. As indicated in earlier sections, the affordable plant-cost penalty for the FBR can be inferred by examining the difference in the ratio for the FBR plant and the thermal reactors. Hence, for U_3O_8 at $40 per lb, the FBR would require a plant cost less than that of the ACR. At a U_3O_8 price of $120 per lb, the FBR plant-cost penalty should be less than 30% to be competitive with the LWR assuming enrichment at $100 per SWU. Assuming an enrichment cost of $50 per SWU, the allowable plant-cost penalty would be only 20%. The situation is even more grim relative to the ACR.

The cost data for the FBR in figure 4.6 assumed only the FBR autonomy strategy. This strategy, in effect, presumes there would be no market for the more valuable U-233 that could be bred in FBR blankets. Figure 4.7 illustrates how the competitiveness might be affected by alternative fuel cycles in the FBR. As can be seen, the situation is improved significantly when some U-233 is bred in FBR blankets.

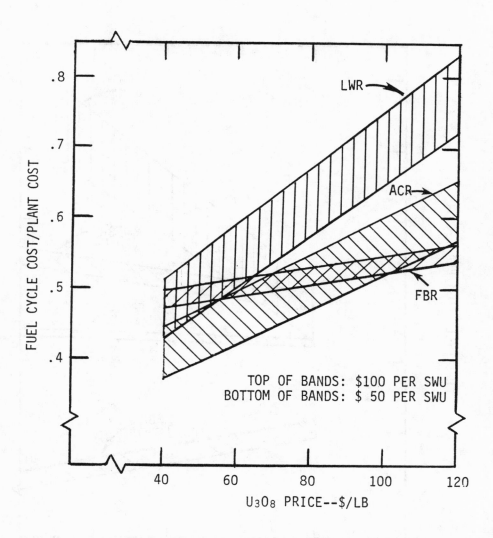

Figure 4.6   (Fuel cycle cost)/(plant cost) ratio as affected by
U₃O₈ price

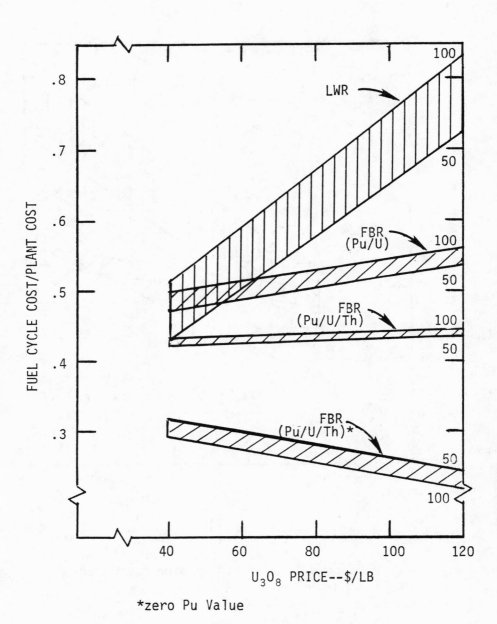

*zero Pu Value

Figure 4.7   (Fuel cycle cost)/(plant cost) ratio for alternative
             FBR plants as affected by U_3O_8 price

One other case is illustrated in figure 4.7. In the very long range, it is conceivable that plutonium fuel could have essentially a zero value, both because of a plutonium sufficiency and a limitation on Pu utilization imposed due to safeguards considerations. If at that time, the ACR user plants should still be willing to pay a price for U-233 based on its competitive value with U-235, then the curve at the bottom of the figure would apply. Obviously, this is a highly-unrealistic case. The value in showing it is simply to illustrate that a lower plutonium value, that might be dictated by limitations on plutonium commerce, could actually favor the competitive position of the FBR. As might be appreciated, this observation has some irony associated with it.

Presumably, in the very long range, the FBR might be justified simply on the basis that it could assure a cheap fuel supply for ACR plants. In that case, some of the cost penalty associated with FBR plants would have to be written off, possibly through relatively large charges on the U-233 fuel to be sold to the ACR customers. Moreover, it is likely that experience with FBR plants will allow some significant cost improvements in their capital costs. Clearly though, the FBR could have a very important role in the long range, if indeed, the price of U_3O_8 increases substantially.

The more important and more immediate concern is how the FBR might be introduced in view of the apparent economic disincentives. One expedient, as illustrated by the data in figure 4.7, is to plan on the use of an FBR fuel cycle that breeds some U-233 for use in thermal-spectrum reactors. This strategy, however, presumes a willing market exists for the purchase of U-233 fuel. In view of these considerations, it is somewhat strange that national policy has put primary emphasis on development of the LMFBR with minimal attention on ACR technology that might provide the market for U-233. This policy appears to put the cart before the horse-- the supply before the demand.

CONCLUSIONS ON THE PREVALENT STRATEGY

The prevalent strategy suffers from a number of problems, primarily associated with near-term considerations. The problems include the following:

1. There appears to be no compelling economic incentive for recycling LWR residual fuels, at least under the current price structures for U_3O_8 and enrichment;

2. The recycle of plutonium also raises questions regarding the possible diversion of weapons-usable fuel;

3. The use of the HEU/Th fuel cycle in presently designed LWR plants is both economically and politically unattractive;

4. The use of the HEU/Th fuel cycle in ACR plants appears to be economically attractive for U_3O_8 prices above $40 per lb, but politically-unattractive because of the use of highly-enriched U-235; and

5. The prevalent LMFBR strategy does not appear to be economically competitive until U_3O_8 prices exceed $120 per lb.

It can be argued that the introduction of the FBR might, however, be justified in some countries having limited indigenous uranium resources and where those countries wish to become more energy-independent. And even in the U.S., continuing evolution of the FBR would seem prudent for the longer range when economically-attractive U_3O_8 resources might ultimately be depleted.

While the nuclear opponent might suggest these disincentives spell a failure for nuclear energy, that conclusion could be extremely reckless. The fact remains that even the present LWR technology is economically competitive with coal in most parts of the country. If and when new surges in our economy occur, it would seem very risky to lean predominantly on fossil fuels once again to satisfy our energy appetite. And, if energy generation in the future must begin to depend on inexhaustible-energy resources, a retreat from nuclear energy could remove the one assured technology option now at our disposal.

Hence, a strong case can be made for a not-so-prevalent strategy that might avoid the apparent problems of the prevalent strategy.

REFERENCES

1.   "Cost-Benefit Analysis of the U.S. Breeder Reactor Program", U.S. Atomic Energy Commission, WASH-1184, January 1972.

2.   Fortescue, Peter, "Assurance of a Durable Nuclear Industry", General Atomic Company, October 1976.

      ------"Association of Breeder and Converter Reactors (A General Picture), Annals of Nuclear Energy, Volume 4, pp. 59-63, 1977.

      ------"Future Tasks for Nuclear Energy", presented at International Scientific Forum on Energy for Developed and Developing Countries, Nice, France, 1979.

      ------"Sustaining an Adequately Safeguarded Nuclear Energy Supply", presented at International Scientific Forum on an Acceptable Future of Nuclear Energy for the World, Ft. Lauderdale, Florida, 1977.

3.   Cohen, Karl, "Implications of Alternative Fuel Cycles for World Economic Development", presented at International Scientific Forum on an Acceptable Nuclear Energy Future of the World, Ft. Lauderdale, Florida, November 1977.

4.   "Nuclear Proliferation and Civilian Nuclear Power", U.S. Department of Energy, DOE/NE-0001, June 1980.

Chapter V

Nuclear Technology Evolution:
A Not-So-Prevalent Strategy

The prevalent strategy outlined in the previous chapter was based on the validity of certain premises regarding energy growth, resource requirements and energy economics. In the light of recent events, those premises have proven to be erroneous. It is necessary, then to recognize those issues that have changed and those that are likely to change in the future as a basis for selecting the most promising technology and institutional courses for development. In particular, the following changes and implications are significant:

(a) energy growth has clearly declined, but could once again accelerate--development courses must recognize both those possibilities;

(b) the economic incentives for bred-plutonium utilization have deteriorated--better tech- nology directions to encourage efficient nuclear-fuel utilization must be sought; and

(c) the public and political acceptance of nuclear energy has fallen into disrepute--better insti- tutional and technology directions to assure energy survival must again be sought.

The first point once again emphasizes the possibility of energy-growth variations or cycles. It was *not* the intent to introduce the subject of energy cycles as a dogma. It *is* the intent, though, to offer a cautionary note that a slowdown in energy growth, i.e., some apparent success in energy conser- vation, should not be interpreted as an end to all future energy growth. Indeed, the assumption of "limits to energy growth" could be a cruel hoax to perpetrate on segments of this country and on much of the world before they have enjoyed the energy-related benefits already available to most of the U.S.

In view of current economic trends for U_3O_8 prices and breeder-reactor capital costs, it will be very tempting to curtail development efforts on more resource-efficient nuclear technologies. And if the world is able to stagger through the oil-supply crisis for another ten to twenty years, it will be tempting to ignore the potential of nuclear energy for solving the portable-fuels problem. But, if a renewed energy-growth surge should occur, beginning some 20 years from now, an unprepared world could resort to devastating means to divide the remaining exhaustible energy resources—in the absence of inexhaustible replacements.

In this context, the cycle-adjusted-logistic energy-growth assumption provides one possible approach for estimating the magnitude of the energy-growth slowdown already underway and the possible resumption of energy growth in another 20 to 40 years. Likewise, a recognition of historical changes in U_3O_8 prices and other economic factors can prepare policy planners for appropriate technology redirections during a period of apparent economic disincentives for the more resource-efficient technologies. This chapter, then, will examine an alternative nuclear strategy during a transitional period between the current nuclear technology and a longer-range, steady-state nuclear technology. In particular, it will seek an appropriate technology-evolution direction that will prepare for a renewed energy-growth surge in the face of economic trends that might not justify the prevalent strategy.

But, of even greater importance is the resolution of problems associated with the public and political acceptance of nuclear energy. As indicated in the previous chapter, at least three previously-unrecognized (or less-appreciated) nuclear technology issues have recently evolved as more critical issues than the earlier recognized ones. These newer issues are:

- a weakening public acceptance of nuclear energy,

- a political concern of eased pathways for
  nuclear-weapons proliferation, and
- the apparent impotence of nuclear technologies
  to alleviate the portable-fuel-supply problem.

In particular, a solution to the first of these problems is crucially important.   Without that solution, the survival of nuclear energy is almost certainly doomed and energy survival itself is very questionable.

The greater urgency for solutions to these previously-unrecognized problems suggests that the more immediate policy planning should be focused on technology and institutional directions aimed toward their resolution. Nevertheless, in the longer range, economic considerations associated with the efficient utilization of bred nuclear fuels must also receive attention. Technology planning even for the latter case cannot be delayed inordinately, simply because a relatively long time is required for the implementation of new technologies.   Both types of problems will, therefore, be discussed in the subsequent sections. First, the newer issues will be examined to determine their impact on preferred technology-development directions. Finally, the economic issues of bred-fuel utilization will be re-examined for a "not-so-prevalent" strategy that could remove some of the current disincentives for resource-efficient nuclear technologies.

TECHNOLOGY DIRECTIONS--THE NEW ISSUES

The importance of energy-generating economics as a factor in selecting appropriate technology-evolution directions is generally recognized. Indeed, if economic benefits could be quantified for factors leading to public acceptance, international stability and a capability for solving our portable-fuel problem, as well as electricity generation itself, the principle of economic selection would have to be regarded as infallible. Unhappily though, an economic quantification of those factors appears to be impossible.

Probably the most urgent problem relative to energy survival is the weakening public enthusiasm for nuclear energy. Possibly the most frustrating problem is the political concern of nuclear-weapons diversion from a technology that may be the only hope for international energy survival. And perhaps the most neglected problem is the lack of attention on nuclear technology as a potential long-range resource for solving the portable-fuel problem.

It has generally been the belief in the nuclear community that solutions to at least the first two less-recognized issues must be primarily educational, political and institutional, i.e., that technology fixes are not possible. Hence, the nuclear technology R&D programs have continued to focus predominantly on the resource utilization and economic issues. While it is, admittedly, quite likely that technology re-directions cannot solve the public-acceptance and non-proliferation direction, it is possible, nevertheless, that narrowly-defined technology directions can preclude some promising institutional solutions. This does not suggest that the established LWR technology directions need be surrendered. But, it does suggest we are not sufficiently prescient to abandon at this time, promising extensions or diversions from the familiar paths.

The need for technology diversity can be illustrated by some preliminary examination of institutional redirections and the effects these potential redirections could have on technology directions. A good beginning point is a re-examination of nuclear growth projections. Recalling that the growth of nuclear capacity was projected to be at least five-fold in the U.S. and ten-fold in the world between the years 2000 and 2030, it would appear that site locations have not yet been committed for some 80 to 90% of the nuclear capacity to be installed in the next 50 years.

With an installed capacity of some 1000 GWe in the year 2030 and 2 GWe per site, approximately 500 nuclear sites would be required in the U.S. In fact, such a policy would lead to nuclear sites only 20 miles apart in some of the more populous areas. Not only would such a policy be unacceptable to most of the public, it would probably be inefficient and not economically desirable. The energy-park concept envisions some 50 to 100 energy-generating centers in the U.S. with, perhaps, 10 to 40 GWe at a single site.

A National Science Foundation study by General Electric [1] in 1975 focused primarily on energy park sizes of 26 GWe. It was their conclusion that the benefits of construction and operation of multi-plant sites did seem to favor the energy-park concept. Probably the most serious problem identified for the use of energy parks was that of rejecting the large amount of waste heat associated with electricity generation in the park. Both the substantial quantities of water required for consumptive use and the potential impact on local climates could place limits on the size and location of energy parks as presently envisioned.

Perhaps an alternative approach to the energy-park concept might be one using a centralized-heat-generation/de-centralized-electricity-generation (CHG/DEG) system. In this case, nuclear power plants would be located in an energy park, but would primarily dispatch heat energy from the park, either as sensible heat or as synthesized gas products. Electricity generation could then be relegated to dispersed sites more conveniently located for electricity distribution and cooling water resources. In this case, waste-heat rejection within the energy park would only be required for the park load itself.

The CHG/DEG concept would, however, require higher-temperature heat than that normally available from LWR plants. This concept would rely primarily on a process-heat reactor for the heat source, probably the HTGR. At least two approaches for transporting process heat have already been studied.

The General Electric company has recently reviewed[2] work being done both in Germany and this country on the use of a thermochemical pipeline (TCP) for energy transport. Above approximately 800°C (1472°F) for the core-outlet gas temperature, heat energy from an HTGR can be utilized to drive a chemical reforming reaction in which CH_4 and H_2O are combined endothermally in the presence of a catalyst to form a synthesis gas mixture consisting of H_2 and CO. These synthesized gases can be transported by pipeline to user locations where the reverse exothermic reaction (methanation) releases the stored energy for process heat or electricity generation.

An alternative approach has been suggested by the General Atomic [3] company using a heat-transfer salt. In this case, the heat energy is transported as sensible heat through a pipeline to a user location.

An attractive feature of either application is that the HTGR technology for generating significantly high temperatures is already established as technically feasible. The use of the molten-salt technique, which might be advantageous for distances up to 50 miles, could require some additional metallurgy development on heat transfer equipment and pipelines. However, some industrial experience with the use of heat-transfer salts already exists. The use of the TCP appears to be more attractive for transport beyond, say, 100 miles. Development work on the reformer and methanator systems is underway in Germany.

The CHG/DEG concept appears to have many attractions that will be discussed in greater detail in the last chapter. A purpose for introducing it here is that this concept is illustrative of the need for carefully selecting promising technologies for continuing support. While high-temperature reactors may not, in themselves, solve the problems of declining public acceptance of nuclear energy and of non-proliferation of nuclear weapons, the technology may more readily permit institutional redirections that could contribute to resolutions of the acceptance problem.

Perhaps the most elusive issue is that associated with the political acceptance of resource-efficient nuclear technologies. The peaceful use of nuclear technology has brought with it a vexing dilemma. Nuclear energy offers to the world, and particularly the large energy-consuming countries, an opportunity for eliminating the extremely serious threat associated with energy-resource dependence on other countries. But a by-product of peaceful nuclear energy is the production of plutonium fuel that could conceivably be diverted to nuclear weapons. To date, much more attention in this country has been given to the weapons-proliferation threat. It is probably not surprising that many of the other large countries, with essentially no indigenous energy resources, are more concerned with the energy-dependence threat. Obviously, both factors are important.

A better perspective on the overall problems of international stability is possible by examining some of the relevant energy-related characteristics of high-energy-consuming (HEC) countries and low-energy-consuming (LEC) countries. In general, the HEC countries:

- make a large impact on the world supplies of energy resources,
- have sufficiently large energy industries to justify energy-efficient technologies whose economics are size-sensitive, and
- are seriously concerned about their energy-dependent vulnerability.

The LEC countries, on the other hand:
- do not yet make a large impact on the world energy-resource supplies,
- cannot yet justify those energy-efficient technologies that tend to be expensive for limited deployment, and
- are more concerned with simply a reasonable growth of energy and its attendant higher standard of living.

A frustration of objectives for either type of countries could lead to political instabilities and the temptation to acquire and possibly use nuclear weapons. The short-term antidote to this problem is to discourage the proliferation of nuclear weapons themselves. The longer-range solution is the elimination of the basic problems.

The current administration (in the U.S.) has been primarily concerned with arresting the potential proliferation of nuclear weapons arising from the greater availability of weapons-grade fuel materials and the facilities to process the fuels. It is correctly pointed out that the economics of fuel recycle is already marginal for the very large volumes of throughput typical of the U.S., and therefore recycle for smaller countries certainly cannot be economically justified. The solution proposed is that the HEC countries should abstain from the implementation of recycle and breeder reactors to discourage the headlong rush toward a plutonium economy. That proposal, perhaps, was a useful shock treatment to bring greater awareness to the risk/benefit ratio of recycle policies, particularly where the limited volume of recycle cannot possibly allow an economic benefit.

As a continuing policy, however, it perpetuates the even-more-frightening prospect of unpreparedness, primarily among HEC countries, to cope with temporary energy-resource shortages or boycotts. Problems of energy-resource assurance are already weighing heavily on many parts of the world. But, the severity of those problems could increase even more dramatically in the event of another energy-growth surge. Based on the history of cyclic energy demand, the world might have only 20 to 30 years to find solutions, other than passive solutions, to the energy-supply problems. Improved energy technologies and institutions should be aggressively sought that can be safely implemented in 20 years and that promise some economic incentive for implementation.

If plutonium is to be used, it is particularly important to develop appropriate technologies and institutions to allow for the safe, secure and economic use of that fuel resource. Early experience in at least a few countries can, and undoubtedly will, supply the leadership in this sensitive industry. The utilization of secure nuclear parks to limit the handling and transport of bred nuclear fuels has been receiving considerable attention as a promising method of enhancing fuel safeguards. The nuclear-park concept for resolving some of the fuel-diversion concerns is consistent, then, with the energy-park concept for resolving some of the public-acceptance concerns. If the concept of an energy center is to serve as a model, though, a "pilot-plant" test of the concept should be undertaken soon, preferably by a multi-national cooperative effort.

A program of this type to minimize the anticipated problems, particularly of the HEC countries, should satisfy at least two objectives, viz:

1. it should move in a direction deemed to be desirable in the long range, and
2. it should utilize appropriate variants of the long-range strategy such that some economic incentive can be assured in the nearer-term transitional period.

The succeeding parts of this chapter will focus on a proposed variant strategy, i.e., the not-so-prevalent strategy, that could, hopefully, satisfy those objectives.

But, before concluding this discussion of international energy concerns, the equally important concern of LEC countries should be addressed. A denial of energy growth to the currently-classified LEC countries could be tantamount to encouraging political instabilities in these nations. The assurance of a productive program for these countries need not involve fuel recycle at an early stage. Indeed, because of the capital-intensive nature of the recycle industry and its economic sensitivity to size, such a program would be economically imprudent for most countries for many years. Unfortunately, though, it has been a prevailing philosophy among HEC countries that nuclear technology developments for their own needs could simply be spun off to LEC countries, thereby providing an export opportunity for the HEC countries. In retrospect, the promotion of this spin-off philosophy has resulted in a serious disservice to the LEC countries.

If the HEC countries are sincerely interested in energy-technology policies to minimize political instabilities in LEC countries, neither technology spin-off nor technology denial would seem appropriate. Instead, the special requirements of these countries should be recognized and cooperative programs should be developed to supply the LEC national needs. Such programs should be neither paternalistic nor exploitive. The LEC countries should be encouraged to play an equal role in planning and participation. If nuclear energy is to contribute importantly in these countries, and it probably should, then technology emphasis should be on:

- smaller reactor sizes,
- very reliable operation,
- safety assurance,
- secure fuel supplies,
- diversion-resistant fuels,

- a once-through fuel cycle, and
- easy waste disposal.

These criteria for technology developments related to LEC-country needs are somewhat tangential to the main subject of the not-so-prevalent strategy—a transitional strategy leading to the more resource-efficient utilization of nuclear fuels. But, this diversion is, perhaps, useful to emphasize the ambiguities in current policy directions that do not properly attack some of the basic problems of both HEC and LEC countries.

Prior to leaving the subject of the less-recognized problems, some additional attention should also be given to the contributions nuclear energy could make toward the replacement of oil. Again, it is probably impossible to project with any certainty what technology directions will ultimately prove most promising to assure a convenient and economic energy substitute for oil and natural gas resources. Perhaps the most important point is that energy institutional arrangements and energy-technology options should be kept open for those directions that could supply some help in areas where oil and natural gas have been used traditionally.

For example, almost 40% of the primary energy consumption in the U.S. is attributable to the use of oil and natural gas in the residential, commercial and industrial sectors. Most of this fluid-fuel consumption is used for low-temperature space heat, water heaters and process steam. In fact, almost 60% of the fluid-fuel consumption in these user sectors is used for low-quality space heating and hot water. The cogeneration of electricity and low-quality heat would seem to offer some opportunity in this application.

Assuming that electricity generation accounts for 30% of our total input energy requirements in the U.S., then an average generation efficiency of about 33% means some 20% of our total energy consumption is rejected as waste heat to produce electricity. If electricity grows to a level that requires 60% of our primary energy input, then more than 35% of our total energy consumption could be rejected as waste heat from electricity generation plants only, even assuming a generation efficiency of 40% for these plants. Several problems make the use of this low-quality waste heat difficult to use, e.g.:

(a) large generating stations generally produce much more low-quality heat than can be used by surrounding communities;

(b) generating stations usually are located away from population centers making the heat distribution uneconomic; and

(c) trends toward still larger and more remote energy parks would amplify the previously-identified problems.

Again, it would seem that the concept of the CHG/DEG systems might improve the potential for utilizing this low-quality waste heat.

Nuclear energy applications to the production of portable fuels should also be sought for the longer range. In the short range, production of hydrogen from methane might be more promising. Subsequently, the production of methane and hydrogen from coal could become attractive. And in the long range, efficient water-splitting into hydrogen and oxygen should be the goal. Probably high-temperature heat from HTGR-type plants will be a key. [4]

In summary, desirable end-use applications of nuclear technology should receive at least as much attention as fuel cycles and breeding. That point, however, is still frequently lost in technology planning.

THE ECONOMICS OF RECYCLE--THE CASE FOR PLUTONIUM

When last heard from in Chapter IV, plutonium utilization appeared to have serious problems, including economic disincentives, diversion concerns and commercialization difficulties. Yet the plutonium resources expected to be stored for the next 20 or 40 years could have a very substantial impact on U_3O_8 requirements during a future energy surge, as was previously discussed in Chapter III. The real challenge, then, to the nuclear industry is to formulate technologies and institutions that can offer both economic and social incentives for its controlled use.

A first requirement responsive to security concerns is the minimization of plutonium handling by recycling this fuel only in plutonium-dedicated reactors, preferably in secure nuclear centers. In these nuclear centers, the reprocessing, refabrication, shipping, fuel storage and utilization of plutonium could all occur under carefully safeguarded conditions. The planning and implementation of these centers is essentially an institutional problem.

It will be assumed, then, that the institutional structures involving primarily the concept of secure nuclear centers can and will be developed to resolve the political problems. Beyond that is the problem of removing the economic disincentives associated with plutonium utilization and actually finding sufficient economic incentives during a transitional nuclear period to assist in commercialization problems. The succeeding discussion, then, will be directed primarily toward the technological and economic factors that could favor such a strategy.

It was noted in Chapter IV that the economics of the FBR improved significantly when part of the uranium blanket was replaced by thorium. Presuming that a market existed for the more valuable U-233, this expedient allowed for the sale of a more valuable bred fuel, thereby offsetting some of the higher costs expected for breeder reactors. The economic potential of U-233 production and a political interest in temporarily attenuating a large plutonium stockpile both suggest that some conversion of thorium to U-233 might be desirable.

Figure 5.1, which is somewhat similar to figure 4.5 of the previous chapter, illustrates the relative fuel-cycle-cost contributions for the extreme case where all of the FBR fertile material--in blankets and core--is thorium. In this role, the fast breeder reactor becomes effectively a fast transmuter reactor. As in figure 4.5, the capital cost penalty for the fast-spectrum reactor has been assumed to be 50% relative to the LWR base case.

The much larger fuel credit resulting from the production of U-233 not only offsets all the working capital interest charge, but also more than half the handling cost. The net relative generating cost is about 1.13, well below the 1.20 established as a target in the previous chapter. Compared to the relative generating cost of 1.38 for the Pu/U fuel cycle, the Pu/Th fuel cycle promises a 15% improvement.

As was pointed out in the previous chapter, a comparison of the fuel-cycle-cost/plant-cost ratio for the alternative fuel cycles is probably a more useful way to illustrate the relative merits. Figure 5.2 shows this ratio for three FBR fuel-cycle alternatives relative to the LWR with self-generated recycle. With the Pu/Th fuel cycle, one could apparently afford a 65% penalty on the capital cost of the FBR plant, assuming U_3O_8 at $120 per lb and enrichment at $100 per SWU. This corresponds to approximately a 40% affordable penalty for the self-sustaining FBR with thorium blankets and roughly a 30% affordable penalty for the conventional Pu/U-238 fuel cycle in the FBR. For enrichment at $50 per SWU, the values are 55%, 30% and 20%, respectively.

Actually, there may be neutronic reasons to use some U-238 in the central part of the FBR core. More realistically,

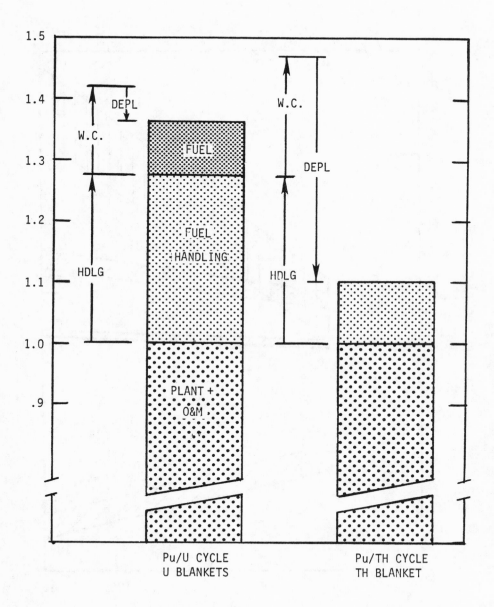

Figure 5.1   Relative generating costs for alternative FBR fuel cycles
(U$_3$O$_8$ at $120 per lb; enrichment at $100 per SWU)

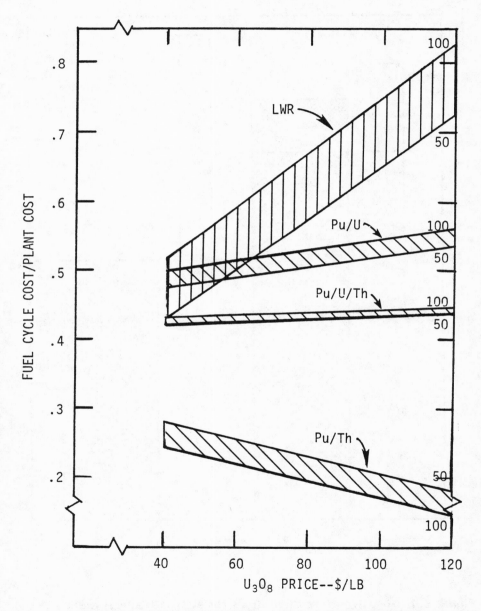

Figure 5.2   (Fuel cycle cost)/(plant cost) for alternative FBR
             fuel cycles

then, the affordable plant capital-cost penalty might lie someplace between 40 and 60%. It is clear, however, that the production and sale of U-233 can provide a very substantial incentive for the predominant use of thorium in the FBR. Again, though, this expedient can only be exploited if a market demand exists for the U-233.

It was also noted in Chapter IV that the economic groundrules for assigning a plutonium value could have a significant effect on the fuel-cycle cost. Figure 5.3 illustrates how changes in the value ratios of Pu/U-235 and U-233/U-235 can affect the fuel-cycle-cost/plant-cost ratio. The top band illustrates the economics for the Pu/U fuel cycle previously illustrated in figure 5.2. Likewise, the intermediate Pu/Th band corresponds to the case shown at the bottom of figure 5.2. Two other cases are illustrated for:

- a reduced U-233/U-235 value, and
- a zero plutonium value.

If the U-233 has a value of 0.6 relative to U-235, then the fuel-cycle-cost/plant-cost ratio is very similar to the Pu/U conventional case where the plutonium has a value of 0.5 relative to U-235. As might be expected, there would be no economic advantage to the use of the thorium cycle in the FBR, if the market value of U-233 is approximately the same as that of plutonium.

If on the other hand, the plutonium has essentially a zero value imposed by its limited commerce, then the FBR using a Pu/Th fuel cycle would have spectacular merits. In this case, the fuel-cycle cost would actually be negative, i.e., the credit for U-233 would exceed all other fuel-cycle cost contributions. For a U_3O_8 price at $120 per lb and enrichment at $100 per SWU, the affordable plant-cost penalty relative to the LWR would be almost 100%. Again, as in the previous chapter, it is emphasized this is a fictitious case, but does illustrate how important bred-fuel accounting policies can be for alternative fuel cycles. More attention will be given to

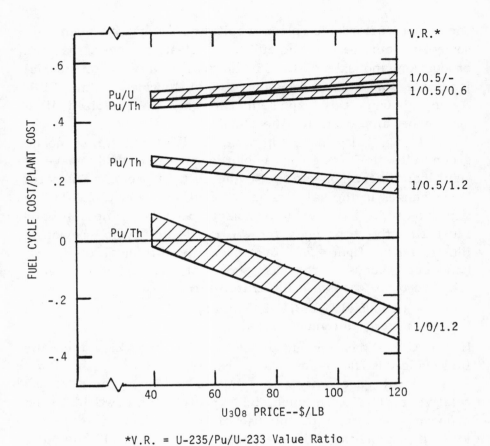

Figure 5.3   (Fuel cycle cost)/(plant cost) as affected by bred
fuel prices

that consideration in the final chapter of this book on institutional expedients.

In summary, then, the use of the Pu/Th fuel cycle in an FBR, or more precisely, a fast-transmuter reactor, can allow significant economic benefits that can be used to offset a high FBR plant capital cost, particularly during its commercial introduction. Clearly, the strategy of burning plutonium to produce U-233 is not a steady-state solution to the quest for a renewable energy source. The plutonium, after all, must be supplied from LWR plants using the LEU fuel cycle, which must depend on a natural source of uranium. But, the strategy appears to be an excellent one for a transitional period when a surplus of plutonium exists and prudence would suggest a modest commitment to FBR plants until more is known about FBR technology, fuel-safeguards and economic problems.

Even this transitional strategy, however, is only practical if a market exists for the U-233. That, then, will be the subject of the next and, perhaps, most important part of the chapter on the not-so-prevalent strategy.

THE ECONOMICS OF RECYCLE--THE CASE FOR U-233

With, perhaps, monotonous regularity, it has been emphasized that the not-so-prevalent strategy can succeed only if a market exists for U-233. It is generally accepted that the use of the thorium cycle and U-233 fuel has considerable attraction for the very long-range future. However, nuclear policymakers, the nuclear-supply industry and the nuclear-user industry have all conditioned themselves to the concept that the development and implementation of the thorium cycle at this time would divert attention and funding from the more important tasks of pursuing plutonium utilization and the breeder economy. Under that dogma, a market for U-233 fuel would not appear to be likely in the next one to two decades.

Clearly, then, the case for U-233 must be re-examined. Development and implementation of the thorium cycle could, after all, be an asset to plutonium utilization, not a diversion. And, concurrent attention on the improvement of resource utilization for thermal-spectrum reactors would not seem to be altogether misplaced in view of the probable preponderance of these reactors for some 30 to 50 years, if not always. It is useful, then, to examine the economic conditions that might make U-233 utilization attractive in the nearer term, particularly for the LWR. Other considerations, such as the relative industrial hygiene and proliferation-resistant characteristics of the alternative fuel cycles will be reserved for a subsequent discussion.

The traditional concept for thorium utilization in thermal-spectrum reactors has assumed a self-generated recycle of U-233 with fuel makeup provided by highly-enriched uranium, i.e., uranium enriched in U-235 to 93%. It will be recalled from Chapter IV that this fuel cycle was economically attractive for the ACR, but not for the traditional LWR. However, the concerns now associated with the commerce of large quantities of U-235 would seem to make the thorium cycle politically unattractive for all cases. Hence, the prevailing strategy for using the self-generated HEU/Th (U-233 recycle with U-235 makeup) fuel cycle in thermal-spectrum reactors appears to have difficulties similar to those of plutonium utilization. While the use of 20% enriched uranium might resolve the political concern, this expedient degrades the characteristics of the thorium fuel cycle sufficiently such that much of the thorium cycle advantages are lost. In summary, both the economics and political concerns make the self-generated HEU/Th cycle somewhat less than a panacea.

One very interesting alternative to the self-generated recycle of U-233 with *HEU* makeup fuel is the recycle of U-233 with *U-233* makeup fuel supplied from an external source. First, the high gamma activity of U-233 makes full decontamination of this fuel less necessary for civilian reactor use--a feature that has some interesting consequences to be discussed later. But, perhaps, more important is a subtle economic distinction between the fuel-cycle costs of the HEU/Th and the U-233/Th cycles as affected by the price value assigned to the U-233. For the HEU/Th fuel cycle, a lower assigned price value to U-233 reduces the fuel-cycle cost by decreasing the working-capital interest charges associated with the recycle inventory of the U-233--typically the U-233 accounts for some 50 to 70% of the total fissile inventory. For the U-233/Th fuel cycle, on the other hand, a lower assigned value to U-233 reduces the fuel cycle cost much more dramatically, partly because the working-capital interest against the entire fuel inventory is affected, but even more importantly, the depletion charges (which now are applied to the U-233) are also decreased. That distinction is very important.

The effect is illustrated in figure 5.4 for the two LWR thorium cycles, first assuming U_3O_8 at $40 per lb and enrichment at $100 per SWU. In this figure, the fuel-cycle-cost/plant-cost ratio is shown as it is affected by the U-233 price value, or more precisely, by the U-233/U-235 value ratio. The cost results are illustrated by bands, with the top of the band indicating fuel-cycle costs based on a refabrication penalty for LWR fuel of six times that of fresh LEU fuel. The bottom of the band reflects a refabrication cost improvement, such that the penalty is reduced to three times. A horizontal band is also shown for the LEU fuel cycle cost assuming a once-through fuel cycle (which is approximately equivalent to the full recycle case).

Some further explanation of the figure is probably appropriate. Again, it will be recalled that the difference between $(FCC)_{LWR}/(PC)_{LWR}$ and $(FCC)_x/(PC)_{LWR}$ represents the fractional plant-cost increase that can be afforded by the system x. Hence, all points above the horizontal LEU once-through line would actually require a lower plant cost for the newer system. Assuming no change in the LWR plant cost for an LWR using the thorium fuel cycle, it is apparent that the user could not afford to pay a value for U-233 exceeding, at most, about 90% of the U-235 value.

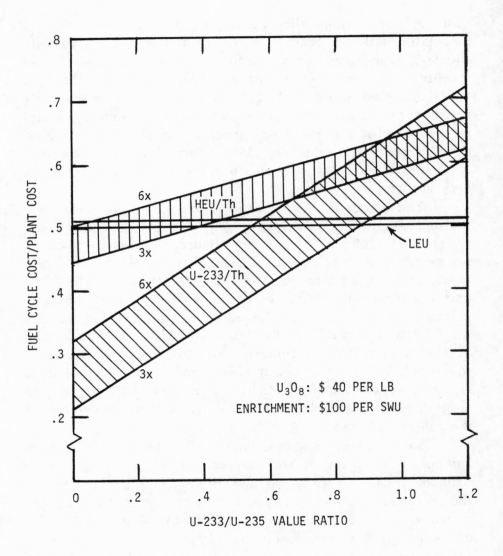

Figure 5.4   LWR thorium fuel cycle cost as affected by U–233/U–235
              value ratio (with fabrication penalty as a parameter)

The relative slopes of the bands show the better economic leverage when U-233 is used as the makeup fuel. However, even with U-233 as the makeup fuel, the affordable U-233/U-235 value ratio would be in the range 0.5 to 0.9, the exact value depending on the refabrication cost penalty. But, with U-235 used as a feed-fuel makeup, the affordable U-233/U-235 value ratio would be only 0.1 to 0.4.

With a U-233 value ranging between 0.5 and 0.9, the economic incentive to the fast transmuter reactor would be less than desirable. One or more of four directions would be required to improve this, viz:

- the LWR might be redesigned to use the thorium cycle more efficiently;
- an alternative ACR, such as the HTGR, could be introduced;
- the refabrication cost for producing U-233 (and recycled U-235) might be improved; and/or
- a price increase in U_3O_8 could be anticipated.

While not shown here, a redesigned LWR or an HTGR could assure an affordable U-233 price significantly above that of U-235, even for U_3O_8 at $40 per lb. That, obviously, would be the most important direction to follow.

The effect of a higher U_3O_8 price of $120 per lb is illustrated in figure 5.5, again for an unmodified LWR. For U-233 feed fuel, from an external source, a price of 90% to 110% of the U-235 price could be afforded. But again, the affordable U-233 price could be increased significantly by modifying the LWR design to use the U-233/Th fuel cycle more effectively and/or by introducing other more resource efficient ACR plants.

Although the light water breeder reactor, being designed by the Naval Reactors organization, has as an objective a conversion ratio (or breeding ratio) slightly greater than unity, a more economically optimized version

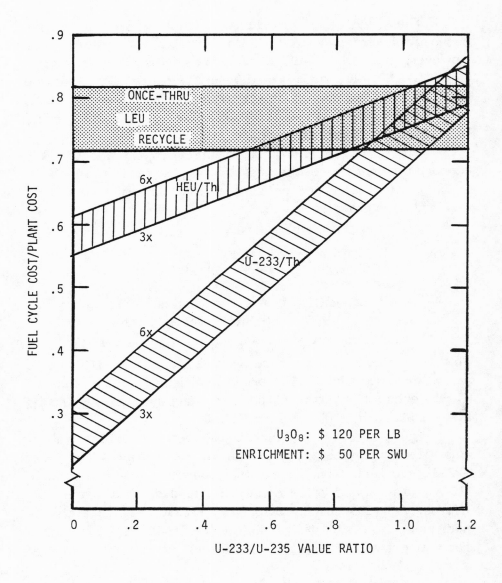

Figure 5.5   LWR thorium fuel cycle cost as affected by U-233/U-235
value ratio (with fabrication penalty as a parameter)

would have a conversion ratio around 0.9. Such a reactor would probably be very attractive for the application outlined here. Additionally, the HTGR could fit the assigned role. The important conclusion is that more attention should be given to ACR plants and the thorium fuel cycle as an expedient for bridging the transitional period when FBR plants might otherwise be discouraged by their apparent high capital costs. Even in the longer range, the ACR plants would continue to be useful as the FBR plants phased into a more traditional breeding role, with blanket-produced U-233 being supplied to satellite ACR plants.

While the not-so-prevalent strategy, then, appears to offer some economic incentives for improved resource utilization, the political concerns of using weapons-sensitive fuels in a civilian industry still require attention. The need for using highly-enriched U-235 has been eliminated by the proposed approach, but Pu and U-233 would still be required. Moreover, some objections relative to plutonium toxicity have not been addressed. Those subjects must be examined.

THE POLITICS OF RECYCLE--THE CASE FOR Pu/U-233

The plutonium economy has met political resistance on two fronts, viz, the toxicity hazards of plutonium and the possible diversion of nuclear fuels for weapons use. Both of those problems will undoubtedly have to be resolved primarily through careful controls, probably including special institutional arrangements. In fact, one of the strongest arguments for energy parks is the improved materials control, accountability and security afforded by a park. Again, those subjects will be discussed in greater detail in the final chapter. Here, the discussion will focus on technological aspects of the toxicity and weapons-proliferation concerns.

Health hazards associated with the handling of radio-active materials can arise from:
- whole-body exposure,
- ingestion (gastro-intestinal tract), and
- inhalation (pulmonary tract).

Gamma radiation usually is a hazard to the whole body while alpha radiation is generally an ingestion or inhalation hazard. The most serious concern with plutonium has been the inhalation hazard. Two excellent reviews [5,6] of the plutonium toxicity problems conclude that the difficulties of handling plutonium have been greatly exaggerated. There has been considerable experience with the handling of plutonium, particularly in the weapons laboratories for some 25 years. With appropriate controls, there is good reason to believe plutonium can also be handled carefully in a civilian reactor industry.

The inhalation hazard of reactor-grade U-233 is some 500 times less serious than that of reactor-grade plutonium, [7] although its ingestion toxicity is roughly similar. Here, the greater concern is simply the overall gamma-radiation exposure that could come from contact operations and handling. Reactor-bred U-233 generally has about 1000 ppm U-232 as a heavy-metal fuel contaminant. Some of the daughter products in the U-232 decay chain have relatively strong *gamma* radiations associated with their decay. The strong gamma-radiation field makes it necessary to utilize remote operations and heavy shielding for U-233 process lines. But, the high gamma activity does actually have some indirect benefits. Due to its strong radiation, contamination from this fuel is more easily identifiable and easily locatable.

The refabrication of plutonium-bearing reactor fuels requires the use of airtight hoods to avoid the inhalation possibility. The handling of plutonium fuel, therefore, is complicated by glove-box operations, which generally makes the cost of plutonium fuel refabrication some three times as expensive as that of fresh LEU fuel. In contrast, the refabrication of U-233 fuel would require shielded facilities and remotely-controlled manufacturing operations as previously indicated. For LWR fuels where a large number of mechanical operations are involved, the refabrication cost penalty using remotely-controlled equipment tends to be five to six times larger than that of the LEU fuel.

Somewhat interestingly, the situation is different for HTGR fuels since in this case many of the operations, such as the cladding of the fuel particles with pyrocarbon coatings, are done by chemical-engineering operations. These operations are amenable to automation and remote control so that the refabrication cost penalty associated with U-233 fuels is only two to three times that of the fresh fuels.

Highly-enriched U-235, U-233 and reactor-grade plutonium are all potentially weapons-usable materials. Once-through, low-enriched-uranium and 20%-enriched-uranium/thorium cycles are generally regarded as acceptable fuel cycles relative to non-proliferation standards. That philosophy would seem to be reasonable with U_3O_8 at a price less than, say, $50 per lb and with reprocessing/refabrication costs at the apparent 1980 level. In fact, as previously discussed, the once-through fuel cycles would also appear to have greater economic attraction for those countries initiating a civilian-nuclear-power program, before the installed capacity reaches at least 20 GWe.

However, a continuing policy of a once-through fuel cycle for the high-energy-consuming countries will only accelerate the demand for U_3O_8, probably forcing prices to levels making the once-through cycle less attractive even for the low-energy-consuming countries. Hence, a continuing emphasis on the passive policy of discouraging improved fuel utilization could actually be counterproductive.

As indicated previously, the longer-range resolution of the problem will probably require institutional initiatives--initiatives that should already be receiving some careful thought.   In this respect, a proliferation-resistant nuclear industry may share the same solutions already required for greater public acceptance of nuclear energy. And again, the desirable institutional redirections might be eased by appropriate technology redirections, and vice versa.

Almost certainly, an important feature of future institutional planning will be the creation of dedicated nuclear parks. Not only will there be merits in concentrating power plants in these centers, but there might be even greater advantages in locating the fuel service facilities in the same reservations.   On the one hand, such a park should be large enough to benefit from the scale of size for fuel service facilities.   On the other hand, the park should be small enough to avoid heat rejection and electricity distribution problems. Both of those requirements can have implications on energy technology directions.

As previously discussed, the heat rejection and electricity problems might be alleviated by technology directions that allow heat-energy transmission to decentralized electricity-generating stations. And, perhaps, the economy-of-size problems for reprocessing and refabrication facilities might be alleviated by appropriate process simplifications. One such simplification might be particularly achievable for the thorium cycle, due to the already-existing gamma activity associated with U-233 fuel.   That simplification, which will be referred to as fuel reconstitution, will be discussed as an illustrative example.

In principle, the fuel-reconstitution concept is similar to the CIVEX concept, [8] that has been proposed for a more proliferation-resistant FBR fuel cycle. That process assumed the elimination of the solvent-extraction decontamination step, which would have two significant consequences, viz:

- it would result in a product fuel with a high
  radiation level, and
- it would eliminate a process step generally
  required for allowing the handling of weapons.

Typically, a fission-product decontamination level of 10^6 to 10^8 is sought for convenient handling of product fuels, yet a decontamination level of about 10 would probably be sufficient for neutronic reasons. If civilian reactor fuels must ultimately be refabricated in shielded facilities, as is certainly the case for U-233/Th fuels, it would seem that the large decontamination factor, generally convenient for weapons materials, might be an unnecessary requirement for civilian reactor fuels. Two questions then arise:

1. Is the cost decrease from the CIVEX reprocessing step sufficient to offset the cost
   increase that might be required for the
   refabrication of LWR fuel?

2. Can a simpler process for achieving a
   decontamination factor of, say, ten be
   developed for the U-233/Th fuel?

It should be recognized, though, that the emphasis in the suggested reconstitution concept is on a less expensive civilian recycle technology that could allow a more favorable opportunity for institutional solutions--not a concept aimed at a technological fix for non-proliferation, per se.

In summary, the economics, handling difficulties and proliferation concerns all favor the use of Pu and U-233 in secure energy centers. While some of the problems associated with the prevalent strategy were fundamentally economic and might be handled solely by technology redirections, new problems have arisen that apparently can only be resolved by institutional redirections. But, even in this case, the insti- tutional redirections might be eased by certain technology redirections,--or at the very least, keeping some promising technology options open. The technology engineer of the future may, indeed, be required to put much more attention on socio-economic problems relative to the more familiar techno-economic problems.

Before concluding this chapter on the not-so-prevalent strategy, the fundamental distinctions between the prevalent and not-so-prevalent strategies will be summarized:

1. The prevalent strategy focused primarily on technological and economic issues within the constraints of existing institutions; the not-so-prevalent strategy suggests that more attention must be given to technological and economic issues to fit within completely new institutional directions--directions that will be necessary to resolve problems not previously appreciated.

2. The prevalent strategy assumed a nuclear growth and U_3O_8 demands requiring prompt implementation of plutonium recycle and early introduction of breeder reactors; the not-so-prevalent strategy assumes a significant energy growth decline for about two decades, with an acceleration of growth around the year 2000 requiring more efficient utilization of energy resources at that time.

3. The prevalent strategy assumed the economics of plutonium recycle and breeder reactors would fully justify their early commercialization; the not-so-prevalent strategy assumes that large inventories of plutonium will accumulate and the traditional breeder-reactor directions will not justify its introduction for, perhaps, 30 to 50 years.

4. The prevalent strategy assumed the introduction of ACR plants and the thorium cycle would represent a diversion to the more important task of deploying breeder reactors and the plutonium economy; the not-so-prevalent strategy argues that the introduction of the thorium cycle with improved thermal-spectrum reactors to utilize U-233 could actually provide an economic incentive for the earlier introduction of fast-spectrum reactors that would initially be used to augment fuel supplies for the thermal-spectrum reactors.

5. The prevalent strategy assumed an ultimate symbiotic system of FBR and ACR plants; the not-so-prevalent strategy assumes the same end-point, but a different transition.

Perhaps it is also appropriate to note that the last assumption may be wrong for both strategies if the fusion-breeder evolves as an economic fuel factory. In that event, continuing development on the thorium cycle would not be lost.

REFERENCES

1.   "Assessment of Energy Parks vs Dispersed Electric
      Power Generating Facilities", NSF 75-500, Center for
      Energy Systems, General Electric Company, May
      30, 1975.

2.   "Chemical Heat Pipe", C00-2676-1, General Electric
      Corporate Research and Development, February 1978.

3.   Vrable, D.L. and Quade, R.N. "HTS Thermal Storage
      Peaking Plant", ORNL-Sub-4188-2/GA-A14160, April
      1977.

4.   Fortescue, P. and Quade, R.N., "Planning for the Future
      roles of Nuclear Energy", GA-A14370, May 1977.

5.   Cohen, Bernard L., "The Hazards in Plutonium
      Dispersal", Institute for Energy Analysis, Oak Ridge,
      Tennessee, March 1975.

6.   "Plutonium Facts and Inferences", prepared by Electric
      Power Research Institute, Palo Alto, California, EPRI
      EA-430-SY, August 1976.

7.   "A Comparison of the Potential Radiological Impact of
      Recycle 233-U HTGR fuel and LMFBR Plutonium Fuel
      Released to the Environment", Oak Ridge National
      Labroatory, ORNL-TM-4768, January 1975.

8.   Marshall, W. and Starr, C., "CIVEX: The Answer to
      Nuclear Blackmail", Nuclear Engineering International,
      July 1978.

Part Three

INSTITUTIONAL DIRECTIONS

Chapter VI

Institutions and Commercialization:
A Historical Perspective

Already it has been hinted that institutional factors could become as important as technological and economic factors in the commercialization success or failure of future nuclear energy projects. It is probable that the continuing commercialization of nuclear technologies will become increasingly difficult due to:

- larger R&D costs than previously expected for complex technologies;
- larger front-end capital requirements for more capital-intensive technologies;
- greater introduction costs, before break-even economics, than previously expected;
- serious financial risks for new ventures; and
- a continuing lack of clear definition between government, supply-industry and user-industry roles.

Where nuclear energy evolution to date has depended primarily on *technology* development, it is likely that the future evolution of nuclear energy may have to depend equally as much on *institutional* developments. Innovation demonstration and ultimate implementation of alternative institutional patterns are sorely needed as both technologies and institutions become more complex and more critical. One solution, of course, is a retreat to a "softer" society with "softer" technologies. Most people in industrialized countries and probably even more people in neo-industrialized countries would undoubtedly object, however, to such a retreat and its associated poorer standard of living.

The remainder of this book, then, will examine the tough institutional problems that must be faced by the nuclear industry. In this chapter, some history of commercialization successes and failures will be reviewed as a basis for understanding the problems. In the next chapter, some possible new directions will be examined. Admittedly, the discussions will be somewhat superficial, partly because that reflects the state-of-the-art. Perhaps, though, recognition of the problems and the possibility of alternative directions will emphasize the importance of the needed innovations and demonstrations of these institutional rearrangements.

Four gradations of commercialization success/failure stories will be reviewed in this chapter. Although the nuclear navy would not normally be referred to as an exercise in commercialization, nevertheless, it has had almost all the ingredients of commercialization. By almost any standards, the nuclear navy must be regarded as a clear success story. The first section, then, will be devoted to that story and its implications on commercialization.

The second section will deal with the utility LWR commercialization success, albeit an expensive success. Third will come a discussion of the HTGR commercialization failure story, and the lessons from that story. Finally, the LMFBR commercialization efforts will be discussed. While this effort cannot yet be labelled unequivocally as a failure story, the history to date certainly gives every indication of promising that fate, particularly if the prevalent strategy is pursued.

In discussing the nuclear navy as a commercialization success, it might be inferred that more pervasive government control is being suggested as an answer. However, exactly the opposite is intended. For a correct perspective, the nuclear navy should be recognized as a project office representing a customer. True, the navy happens to be supported by government funding. But, the navy is more accurately a *customer* for nuclear energy and, in that role is analogous to the utility industry. It is for that reason that the nuclear navy provides a useful case study.

THE NUCLEAR NAVY SUCCESS STORY

A book on energy intended to be provocative would be incomplete without some discussion of Admiral Rickover and the nuclear navy. Among most scientists, industry leaders and the top echelon of Navy officers, the Admiral would probably score low in a popularity contest. Notwithstanding his popularity rating, it is an undisputable fact that he and the nuclear navy have had a remarkably successful record, both in technology development and in overall nuclear-technology implementation.

The general public would probably associate Admiral Rickover primarily with the technological achievements of nuclear energy. The more important theme to be pursued here, however, is that the nuclear navy's accomplishments relative to organization, institutional arrangements and policy implementation might actually leave a more indelible mark on the nuclear industry than the technological achievements themselves. In effect, the early development efforts were successfully "commercialized" into a multi-project nuclear navy, and it is that achievement that will receive primary attention here. An objective of this review is to examine how and why things went right in this success story. The intent is to search for similar directions that might be pursued by government and industry in other commercialization endeavors. As a preview to that subject, however, it is interesting to review briefly some of the technology achievements of the nuclear navy.

In retrospect, it is remarkable that the technology development and deployment of nuclear systems in the early nuclear submarines could have been accomplished in such a short period of time. While some thought had been given to the concept of a pressurized water reactor at the Oak Ridge

National Laboratory as early as 1946, formal studies only began at the Argonne National Laboratories in 1948. A reactor division was established in the Atomic Energy Commission in January of 1949, which included a nuclear reactor branch, headed by Admiral Rickover (a Captain at that time). Later in the year of 1949, the PWR was selected as the most promising concept to be pursued for a submarine propulsion plant.

It was in the middle of 1949 that Westinghouse was brought into the program formally and the Bettis Laboratory was established. Only four years later, the prototype plant for the Nautilus submarine in Idaho was brought into operation, a truly astounding achievement. And, in less then six years, from project initiation, the Nautilus submarine made its initial sea voyage under nuclear power.

The development work leading to this propulsion plant included, not only the reactor design itself, but the development, design and construction of the steam generator and steam propulsion equipment. Perhaps even more remarkable, the development included the research, testing and qualification of a completely new metallurgy technology involving the use of zirconium for fuel elements; a technology that required the establishment of an entire new production industry.

Almost equally as remarkable was the development of the alternative sodium-cooled reactor system for the Sea Wolf submarine. Here, an agreement was reached with the General Electric Company in the middle of 1950 to use the Knolls Atomic Power Laboratory for this task. In spite of the more complicated system and the well-known incompatibility of sodium and water, a prototype was designed, constructed and operated by the year 1955; i.e., within a period of only five years.

Even in the area of civilian nuclear power, proper credit must be given to the nuclear navy organization. In the middle of 1953, the AEC authorized a civilian power plant to be designed and built at Shippingport, Pennsylvania. Ground breaking for the project occurred in the middle of 1954 and full power operation (60 MWe) was reached by the end of 1957, only three and one-half years from project authorization to commercial operation. Again, this must be recognized as a remarkable achievement in view of the fact that development, design and acquisition of all components were involved. In this particular case, the development included a zirconium-clad, uranium-oxide fuel element that was used for the blanket fuel.

Many factors were involved in the success of these projects, including the overall organization, the technical direction and the program support activities. A theme that is frequently belabored in historical discussions of nuclear navy projects is the conflict that developed between engineering and science. The early struggles between Rickover and the laboratories were particularly famous for the conflicts between the science-oriented staffs of the laboratories and the engineering orientation that Rickover sought. These conflicts are well documented, for example, in the book, Nuclear Navy. [1] There is no doubt that the laboratories in the late 1940s were, indeed, heavily dominated by programs aimed at the development of fundamental nuclear data and the evaluation of the relative characteristics of alternative reactor design directions. It would be unfair, however, to suggest that this work was unnecessary. And, for example, many outstanding physics, metallurgy and engineering contributions came from the KAPL program even before the nuclear navy assignment.

It is, though, a credit to the nuclear navy organization that they were able to orient the laboratories toward successful project organizations when much was yet to be learned about the fundamentals of nuclear technology. To allege, however, that Rickover was not sympathetic to fundamental scientific and engineering activities, where they were important, would be grossly unfair. The naval reactor group was a strong champion of careful critical experiments, sophisticated nuclear reactor theory, the use of digital computers for reactor physics calculations, an understanding of the irradiation behavior of fuel materials, the development of shielding theory and supporting experiments, and many fundamental engineering studies, including elaborate heat transfer experiments. In fact, both Bettis and KAPL became spawning grounds for much of the reactor technology that was developed in this country.

But one distinction was eminently clear, the development work as well as design work was intensely mission-oriented and was driven almost fanatically by schedule pressures. A new breed of physicists, metallurgists, chemists, mathematicians and even engineers evolved from the discipline of the nuclear navy program. The new breed was characterized by a dedication to the solution of practical problems and a team camaraderie that focused on achieving goals clearly defined as program requirements. In this respect, then, the science as well as the engineering associated with reactor development, had changed its complexion.

While these technology achievements were very impressive, the organizational achievements leading from an initial development program to demonstration and ultimately to a full-scale nuclear-navy implementation were even more impressive. In the course of this organizational evolution, the Naval Reactors branch extended its management from the initial role of development programs to include:

- project management relative to vendors,
- project management within the navy shipyards,
- assurance of strong manufacturing organizations,
- adoption and enforcement of strong quality assurance standards, and
- operator selection and training.

In one sense, the NR (Naval Reactor) role in the AEC was somewhat accidental. One of the provisions of the 1946 Atomic Energy Act was that atomic energy development must be administered through the Atomic Energy Commission under civilian control. Probably partly because of a post-war reaction to military domination of atomic energy development, priorities in the AEC and, particularly, the AEC laboratories tended to favor civilian applications at some expense to navy interests. To solve this problem, NR was given a dual responsibility, one reporting to the navy and the other to the Division of Reactor Development in the AEC. This dual responsibility allowed considerable flexibility for NR, particularly where administrative red tape was involved. The important point was, though, that NR was basically a representative of the navy customer, not a director of R&D within the AEC. That distinction is important, because it gave NR the motivation and authority of the user--a much more powerful role than that of an R&D director.

Rickover put strong emphasis on the role of NR as a customer. He staffed NR with a relatively small, but carefully selected and trained group of technical people who were charged with the responsibility to review, understand and guide all programmatic work essential to the success of the projects. The emphasis throughout NR was on technical management, not administrative management. He was an adamant believer in thorough training, not only for his own staff and related navy staffs, but also within the vendor organizations. It was largely through his initiative that nuclear training schools were established at Oak Ridge National Laboratory and at Bettis.

The normal government method of dealing with industry contractors was to delegate responsibilities for the administration and surveillance of contracts to staff people. The NR method was one of total immersion. Where important decisions were necessary, appropriate NR staff people frequently participated in the meetings where those discussions were forthcoming. Every attempt was made to assure that technical decisions were ultimately the responsibility of the vendors, but the pressure of NR interests was not an insignificant factor.

On the basis of considerable navy experience with procurement of high-technology equipment from suppliers, the role of experienced industry was recognized as an essential ingredient for success. Hence, rather than attempting to build its own technical development and design organization, Rickover pushed aggressively to involve Westinghouse and General Electric. But, because of his own strong interest in assuring that the customer was satisfied, he involved himself in company management organizations and policy activities, more than was popular. It is interesting that throughout the NR dominance of industry management, the size of the industry staffs was generally well controlled.

The important observation from this review, though, is that technical excellence was regarded as the hallmark of success, and that excellence was expected through all management levels to the top level. High technology was recognized as the product, and high-technology management was regarded as the key.

As the nuclear navy program moved from single projects to multi-projects, NR recognized the need for greater discipline and organization in the manufacturing areas. The responsibilities of Bettis and KAPL were defined primarily as nuclear systems development and specification. Separate manufacturing organizations were established in Westinghouse and General Electric.

The role of NR in pioneering strong quality assurance programs in the nuclear industry is well known. The selection and training of operational crews was also recognized as a critically important responsibility. While all of these activities were important and were accomplished effectively, an underlying key to success was the recognition and definition of the roles of the AEC, the NR customer representative, the navy customer, the development and design centers and the equipment suppliers.

In summary, the nuclear navy was able to foresee and plan the appropriate organizational and procedural steps to move from a development organization´ to a fully-deployed nuclear navy. And they succeeded.

THE LWR QUALIFIED SUCCESS STORY

In 1953, the prototype plant for the Nautilus submarine was brought to full-power operation. Even prior to that date, naval reactors had initiated a study on a land-based prototype for an aircraft-carrier power plant. Largely as a result of national budget economies, this project was cancelled in 1953. Simultaneously, though, there was strong interest within the congressional Joint Committee on Atomic Energy to undertake a national effort directed toward a civilian nuclear power plant. Amidst considerable controversies over military/civilian, private/public and industry/navy political concerns,[1] it was concluded that a pressurized-water-reactor project, based on the aircraft-carrier preliminary studies, should be pursued as part of an "Atoms for Peace" program.

Hence, in 1953, the Shippingport PWR project was authorized. In 1957, the Shippingport reactor was brought to full-power operation. The outstanding success of that project can be attributed to a number of factors, including the already-well-understood PWR technology, the involvement of an experienced reactor design team and the disciplined project management of the NR team. The Shippingport PWR program was clearly an important initial step in the LWR commercialization success.

In 1955, the AEC announced a Power Reactor Demonstration Program (PRDP) designed to bring private industry more actively into the national reactor development effort. Over a period of several years, a total of some 12 reactor projects were committed under this program. One particularly important project, relative to LWR commercialization, was the Yankee PWR project involving 12 northeastern utilities and the Westinghouse Corporation. Perhaps somewhat more surprisingly, General Electric committed to the development and construction of a boiling water reactor (BWR) plant outside the PRDP program and completed the Dresden BWR one year ahead of the Yankee PWR.

Both vendors and utilities recognized the nuclear commercialization efforts were not without financial risk, though the degree of risk was probably much greater than anticipated in the 1950s. Indeed, it is likely that some of the nuclear vendors would not have pursued even the LWR development if hindsight could have been converted to foresight. Recognizing that some forward investments and subsequent risks were involved, a variety of contractual arrangements were devised to share risks between vendors, utilities and the government. The cost/risk arrangements for four types of approaches are illustrated in table 6.1. The approaches are identified primarily by typical examples for each arrangement.

In the table, the risk to the vendor increases in descending order of the examples. Hence, the vendor risk assumption was minimal, if not zero, for the Shippingport project, and was maximal for the Fort St. Vrain approach. It is interesting that Westinghouse pursued the best-known technology, i.e., the PWR, but chose to lean heavily on the PRDP program in early years. In contrast, General Electric chose the more venturesome BWR technology, yet selected the turnkey approach. Both temporarily used turnkey contracts subsequently for plants with higher power ratings. And it is generally recognized that both vendors absorbed substantial financial losses in their commercialization ventures.

Table 6.1   Examples of Cost/Risk Sharing Arrangements
Used in Typical Approaches

Approach Typified By:	Cost/Risk Responsibility			
	Development & Design Cost	Construction Base Cost	Construction & Performance Risk	Availability Risk
Shipping-port	gov't	gov't/utilities	gov't/utilities	utilities
Yankee	vendor/gov't	utilities	utilities/vendor	utilities
Turnkey	vendor	utilities	vendor	utilities
Fort St. Vrain	vendor/gov't	utilities	vendor	vendor

In summary, then, the LWR commercialization can be classified as a success story from the point of view that large numbers of reactors were subsequently marketed and constructed. This particular success of the LWR can be attributed to a number of factors, including:

1. the technology was less complex than most of the others,
2. the technology evolved from a strong naval reactor base,
3. strong engineering teams were available to the vendors,
4. the PWR and BWR plants were introduced primarily by strong and sophisticated utilities,

5. the government initiative in the Shippingport
    and PRDP programs created important
    momentum, and
6. the climate for nuclear technology uti-
    lization was favorable.

In spite of the success story for the LWR commercialization, it is likely it could not be repeated by similar
projects in the future. Indeed, it is even questionable that the
large vendors who introduced the LWR would have pursued
that technology under the existing institutional arrangements
had they known then what is now known. Hence, the LWR
commercialization could be described more accurately as a
qualified success story.

THE HTGR FAILURE STORY

Still another gradation lower in the spectrum of commercialization successes is the attempted HTGR commercialization.
Its importance is that it is very likely to be a harbinger of
similar experiences, if the traditional institutional approaches
are not modified.

While the evolution of HTGR technology has seen more
than its fair share of problems, the fundamental technology
still appears to be sound. Indeed, some of its unique
application opportunities, its resource efficiency and its
different safety features continue to be recognized as
benefits that should be pursued.

The fixed-price contracts negotiated for the large
HTGRs marketed in the 1971-73 period are frequently
identified as the cause of the HTGR commercialization
demise. Those contracts were certainly a factor, though the
seeds for the failure probably were sown by the Fort St. Vrain
contract that was executed some seven years earlier. To understand better how the HTGR commercialization was aborted, it
is useful to trace some of its history.

In 1959, General Atomic, then a division of the General
Dynamics Corporation, signed a contract with a group of 53
utilities, the High Temperature Reactor Development Associates, to develop and build a 40 MWe HTGR plant on the grid
of the Philadelphia Electric company. The contract, which was
part of a third round of the AEC PRDP program, established
a fixed price of $24.5 million to the utilities and committed
the AEC to a contribution of $17 million, including a waiver
of use charges on the fuel. The plant achieved full-power
operation in 1967, though not without considerable construction
pains—including a fire that delayed completion of the plant
about one year. The first fuel charge suffered some
irradiation-induced deterioration, requiring its replacement at
about half its scheduled performance life. The replacement
fuel with improved coated particles, however, performed well
for the entire specified fuel-life. After some seven years of
operation, the plant was decommissioned, primarily because it
had served its purpose very well and was too small to be
economically attractive for continuing operations.

By all standards, it probably would be agreed that the
Peach Bottom HTGR feasibility experiment was an outstanding
success. The success could be attributed to a number of
factors, including:

1. a technology with no basic flaws,
2. a strong development and design team led
   by a brilliant and thorough project manager,
3. a technically strong utility, and
4. a reasonably good AEC/utility/supplier
   contractual agreement.

In 1965, the Public Service Company of Colorado con-
tracted with the General Atomic* company for the construction
of a 330 MWe HTGR plant at the Fort St. Vrain site in
Colorado. The contract again was a three-way contract with

*GA was still a division of General Dynamics in 1965, but was ac-
quired by the Gulf Oil Company in 1967, with 50% subsequently ac-
quired by the Royal Dutch-Shell group in 1973.

the AEC, a utility and the supplier. In this case, though, the contract included an availability warranty. It was agreed by the supplier that a gas turbine plant and its fuel supply would be provided to PSC if the nuclear plant failed to meet its schedule for a 1972 startup. Another feature of the contract was that GA would assume the responsibility for the nuclear fuel supply, with AEC assistance, and would collect revenue from PSC at a fixed rate per kilowatt-hour of electricity generated. Although the fuel contract allowed for some escalation provisions, the escalation was based on normal price indices, thereby giving little protection for abnormal cost increases associated with U_3O_8, separative work and fuel fabrication.

In 1977, the FSV reactor completed its pre-operational test program and started the ascent to power. Difficulties were encountered with various plant components and with a core-temperature instability. The latter problem resulted in local core temperature fluctuations, under certain operating conditions, precluding generation above 70% of rated power until the problem could be rectified. In general, the plant and reactor problems encountered were typical of those that might be expected in a demonstration plant. But, the contract conditions allowed little, if any, flexibility for risk sharing between the three parties. As a result, the financial burden on General Atomic became overwhelming.

In 1979, an agreement was reached between General Atomic and Public Service of Colorado whereby the plant would be turned over to PSC for a settlement based on a derated plant. It is likely that the total GA investment in this demonstration effort reached at least $400 million, including contract settlement commitments.

Many of the problems that arose could probably be attributed to a lean R&D program in the face of major design changes from the Peach Bottom plant to the Fort St. Vrain plant. Some very substantial problems also developed from

changes imposed by redirections of the regulatory agency. The
intent here is not to analyze all the problems and their causes,
however. Suffice it to say that problems can be expected in
the introduction of any new technology, and the severity of
the risks can be affected by:
- the degree of technology innovation,
- the R&D (or lack of R&D) prior to demonstration and introduction,
- the stability of licensing requirements,
- the size (and, therefore, the financial
  investment) of the plant, and
- the degree to which the risks can be shared.

While the HTGR technology was not substantially more
complex than that of the LWR, it did not have the strong
technology base that was available to the LWR from the
nuclear-navy program. Even more importantly, the AEC during
this period was preoccupied with the development of the
LMFBR and, therefore, very little funding was accorded the
HTGR development. In general, the development funding for
fuel testing exceeded that for all component testing, and
funding for systems testing (such as the steam system or the
circulator bearing-water system) was minimal. Still another
factor could be that the General Atomic Company was the only
gas-cooled-reactor contractor in the U.S. In this position, some
government people may have viewed R&D support for
gas-cooled reactors as support to a company rather than a
technology.

The limited technology base, the open-ended contractual
commitments to the vendor and the depressing sequence of
troubles laid the groundwork for the commercialization failure
of the HTGR. The subsequent contracts for large HTGR plants
only sealed the doom of the HTGR commercialization attempt.

In 1971, Philadelphia Electric agreed to the purchase of
two 1160 MWe HTGR plants and later that year Delmarva
Power and Light announced plans to purchase two 770 MWe

HTGR plants. In the following two years, contracts for four additional plants were signed and at least two additional plant commitments had reached the planning stage. While these contracts did not warrant availability, as did the Fort St. Vrain contract, it was necessary for a newcomer in the nuclear industry to commit to turnkey contracts for the nuclear steam supply system and some ten years of fuel supply. Contract prices were based on the assumption that a substantial HTGR business would evolve, so that somewhat higher costs associated with earlier plants might be balanced by some profits from subsequent plants.

Disaster visited this venture in at least three ways:
- continuing problems with the Fort St. Vrain
  project forced second looks,
- the cost escalations associated with high-
  technology construction [2] greatly exceeded the
  escalation of price indices in general (which were
  the basis for price adjustments in the contracts),
  and
- utility plans for expansion were severely curtailed.

For the strong believers in economic cycles, it might be concluded that the General Atomic venture was the victim of timing. To some extent, this may be true, but it is doubtful that commercialization troubles could have been avoided even in a high-growth period, assuming the traditional institutional directions for technology introduction.

In 1973, contract cancellations in the nuclear industry began to predominate over contract commitments. General Atomic, for understandable reasons, became a leader in this dubious distinction. With their commitments reduced to four plants and with the Fort St. Vrain contract as a constant reminder of the hazards of open-ended contracts, the owners of General Atomic terminated the remaining contracts at some further large expense. Altogether, it is probable that the total private investment in HTGR technology has approached a level of a billion dollars, including contract termination costs. Of course, much of that cost was associated with the Fort St. Vrain demonstration program.

Again, it is emphasized that this account of General Atomic difficulties with the HTGR commercialization venture is not intended to be a detailed analysis of technological, contractual and management problems. The important point is that the HTGR commercialization venture went a step beyond that of the LWR, both in technological innovation and contractual commitments. In view of the difficulties already experienced by the LWR vendors, even where a more solid technology base existed from naval reactors, it is not surprising that the HTGR commercialization was a failure story. Probably technology risks associated with a new reactor concept could be reduced significantly by greater R&D planning and support on a national scale, but commercialization problems would undoubtedly continue to plague the industry without substantial changes in the institutional arrangements for sharing costs and risks. Indeed, without those changes, it is doubtful any large-scale, new technology will again be introduced to the marketplace.

THE POSSIBLE LMFBR FAILURE STORY

Based on both the LWR and HTGR commercialization experience, it must be concluded that the commercialization of the LMFBR is likely to be a failure story without changes both in the transitional technology direction and institutional arrangements, even though heavy government subsidies could relieve some of the financial burdens of R&D and demonstration plants. Hence, this particular discussion is labeled a "possible failure story". However, if the problems are properly recognized, and if alternative technological and institutional approaches are properly pursued, it is possible that an LMFBR success story could result--though that possibility looks rather grim at this time.

Beyond the economic problems associated with the LMFBR commercialization itself, the political problems of introducing a large plutonium (or U-233) economy would also appear to be substantial. The transport of plutonium fuel from a number of fuel service centers to an even larger number of utilization centers may be resisted politically at local, national and international levels. Persuading industry to re-enter the fuel processing business will be very difficult after the U.S. government policy of a reprocessing moratorium, costing the Allied/General-Atomic Nuclear Services hundreds of millions of dollars. And politics within executive and legislative branches of the government will make it equally as difficult for the U.S. government to enter this business.

Consolidation of fuel handling services in a few nuclear parks might solve some of the handling problems associated with sensitive nuclear fuels. But, plutonium transport to dispersed generating sites could still be controversial. Perhaps that problem could be minimized or eliminated by locating most of the reactors using bred fuel in the nuclear parks. This, however, might impose limitations on utilities that could create serious electricity-distribution complications and added electricity service expenses. And private ownership of large nuclear complexes, or even shared ownership of the facilities could create still other problems. It is virtually certain, then, that the institutional patterns as they now exist will not suffice for the large-scale introduction of the LMFBR and plutonium utilization.

Though the political problems associated with the commercialization of the LMFBR and its fuel services would appear to be severe, the economic problems of cost and risk may be even more formidable. As indicated in the discussion of the prevalent strategy, the capital cost of the LMFBR must be less than 30% above that of the LWR assuming:

- U_3O_8 at $120 per lb,
- separative work at $100 per SWU, and
- a market for blanket-bred U-233.

Under the more likely assumption of:
- U_3O_8 less than $120 per lb,
- separative work at $50 per SWU, and
- no market for U-233 (assuming no attention
  on the thorium cycle),

the capital-cost penalty of the LMFBR relative to the LWR would have to be less than 15% (see figure 5.2).

Even if the economic conditions should be favorable relative to cost competiveness of the 20th or 30th FBR plant, it would still be necessary to cover the incrementally higher costs of early plants, i.e., the costs before significant learning. These cost penalties and the additional cost risks associated with design and construction are particularly difficult to quantify. However, some general feeling for average commercialization costs over several plants can be afforded by an examination of possible cost reductions based on classical learning theory.

Empirical learning theory assumes that the cost of equipment decreases a fixed percent each time cumulative production is doubled. For example, the cost of a second piece of equipment would be, say, 10% less than the first. Similarly the cost would decrease:

- 10% for the fourth unit relative to the second,
- 10% for the eighth unit relative to the fourth,
- 10% for the sixteenth unit relative to the
  eighth, etc.

The Rand Corporation [3,4] has used the "learning-curve" methodology in the study of costs for military equipment and has recommended its greater use for reactor technology planning.

In general, research on learning-curve methodology has indicated that learning rates typically fall in the range of 80% to 90% (or 10% to 20% incremental improvement each doubling of production). The illustrative analysis here will assume that the learning rate for the LMFBR capital equipment will, on the average, follow an 85% learning curve. The usual procedure for utilizing learning curves is to estimate the cost of the first or second plant as a reference and, thence, calculate the number of plants required to reach a competitive level. Here, it will be assumed, somewhat arbitrarily, that the 32nd FBR plant will reach an asymptotic value. The learning curve methodology will then be used to calculate the cost increment over the first 31 plants required to establish the commercialization.

If one assumes, for example, that the cost of the 32nd FBR plant should be 1.25 to 1.50 times that of an LWR plant, then a band of learning curves as shown in figure 6.1 can be developed. For a 32nd plant cost of 1.5 times the LWR, the first plant cost would be 3.38 times the LWR plant cost; the second plant cost 2.87 times; the fourth, 2.44 times; the eighth 2.08 times and the sixteenth 1.76 times the LWR cost. For a 32nd plant cost of 1.25 times the LWR, the first plant cost would be 2.82 times the LWR, etc.

If the initial plant cost ratio of the FBR to the LWR is 3.38/1 and the affordable capital-cost-penalty ratio is 1.5/1, then the excess penalty ratio for the first plant is 1.88/1, i.e., the cost penalty for the first plant is 188% of an LWR capital cost--a very substantial cost penalty, indeed. The penalty of the second plant on this basis, would be 137%; the penalty for the fourth plant 94%; for the eighth plant 58% and the sixteenth 26%. Summing the penalty over the first 31 plants, the total penalty would be almost 13 times the cost of an LWR plant, or effectively, an average penalty of about 40% of the LWR cost for each of the first 31 FBR plants. This penalty is, then, the net commercialization cost in addition to the assumed asymptotic 50% penalty. That commercialization cost penalty, of course, does not include other development costs such as R&D, component testing, fuel testing, demonstration plants nor the commercialization cost of a supporting fuel cycle industry.

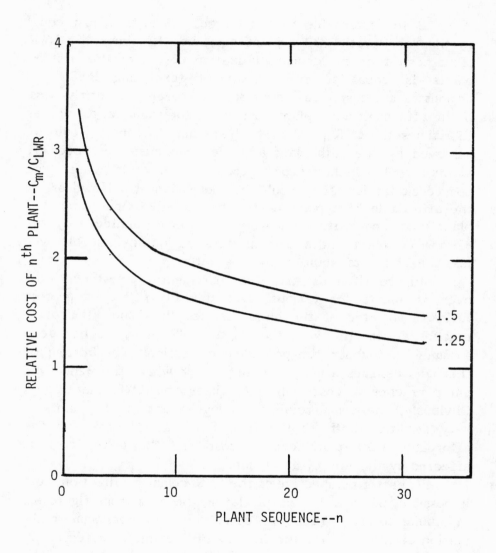

Figure 6.1   Illustrative learning curves for FBR assuming a learning
rate = 0.85

If the affordable FBR plant cost is, in fact, just equal to the "asymptotic" 32nd plant cost, i.e., 1.5 times the LWR plant cost, then the commercialization cost over the first 31 plants is equal to the investment (excluding R&D and demonstration costs) cost necessary to reach competitiveness. If the affordable FBR plant cost were, for example, as high as 1.75 times the LWR plant cost, then competitiveness would be achieved by the 16th plant and the investment cost over 31 plants would include some credits. In this case, the commercialization cost would be more modest. However, if the affordable FBR plant cost were only 1.25 times the LWR plant cost, competitiveness would not yet be achieved by the 32nd FBR plant and the commercialization investment over just the 31 plants would even be greater.

Figure 6.2 illustrates the incremental investment cost over 31 plants for various affordable FBR/LWR plant cost ratios. The top of the band assumes the 32nd FBR plant reaches a cost 1.5 times that of an LWR plant, as has been assumed throughout the preceding discussion. The bottom of the band assumes a more optimistic case where the 32nd FBR plant reaches a cost only 1.25 times the LWR plant cost. Obviously, the commercialization investment for 31 plants is very sensitive both to the final FBR capital cost and the affordable FBR plant capital cost--and the latter cost is affected by the fuel cycle choice.

A better perspective of the FBR commercialization cost is possible, perhaps, by recapitulating the data from figure 6.2 in tabular form. Table 6.2 summarizes the net commercialization investments for the first 31 FBR plants as affected by the "asymptotic" cost (of the 32nd plant) and by the affordable FBR plant cost. If the asymptotic FBR/LWR plant cost ratio is 1.5, then it appears that a fuel cycle should be chosen to allow an affordable FBR/LWR cost ratio around 1.75 to keep the commercialization cost below the equivalent of 5

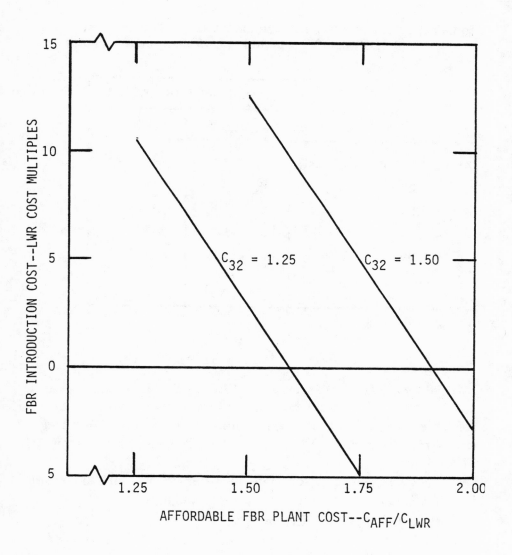

Figure 6.2   FBR introduction cost for 31 plants as affected by the
affordable FBR cost and the asymptotic cost

Table 6.2   Typical commercialization investments in the first
            31 FBR plants measured in LWR plant cost multiples

	"Asymptotic" FBR/LWR Cost	
	1.25	1.50
Affordable FBR/LWR Cost:		
1.25	10.6	--
1.50	2.8	12.8
1.75	-4.9	4.8
2.00	--	-2.8

LWR plants.   If the asymptotic FBR/LWR plant cost ratio is
1.25, then a fuel cycle choice could be made allowing an affordable FBR/LWR cost ratio of about 1.45 without incurring a
commercialization cost of more than 5 LWR plants over 31
FBR plants.

Recalling the discussion in Chapter V, the affordable
FBR/LWR plant cost ratios were very roughly:

1.2 for the Pu/U cycle;
1.3 for the Pu/U/Th cycle;
1.6 for the Pu/Th cycle; and
2.0 for the Pu/Th cycle with zero Pu value.

Apparently, the commercialization cost could be enormous,
then, for the traditional Pu/U fuel cycle, even with an

"asymptotic" FBR/LWR cost ratio of 1.25.

Once again, it is emphasized that the learning-curve approach to identifying probable commercialization costs is only illustrative. Probably the same conclusions could have been apparent from more simple observations. Those conclusions are:

1. It is very important to minimize the capital cost of the FBR;

2. The thorium cycle can be used to ease commercialization costs of the FBR; and

3. Institutional rearrangements will probably be required under the best of circumstances to introduce the FBR.

This, in no way, is intended to minimize the need for breeder reactors in the long range to provide an inexhaustible energy resource. Beyond the transitional period of the next 30 to 50 years, the continuing worldwide growth of energy will place even more severe strains on the supply of energy resources. The price of U_3O_8 (in 1980 dollars) will probably rise to even higher levels than the $120 per lb assumed in this analysis, in which case the affordable FBR plant cost would be even greater, perhaps greater than 2.0 times the LWR cost. And, indeed, a symbiotic combination of FBR and ACR plants would, in turn, combine most of the advantages of low capital costs associated with thermal-spectrum reactors and low fuel costs deriving from breeder reactors.

The basic problem, then, is simply bridging the gap from current technologies and institutions to the asymptotic technologies and institutions. And institutional redirections will certainly be an important requirement.

REFERENCES

1. Hewlett, Richard G. and Duncan, Francis, "Nuclear Navy", University of Chicago Press, 1974.

2. Faltermayer, Edmund, "The Hyperinflation in Plant Construction", Fortune Magazine, November 1975.

3. Letter from Donald B. Rice, President, Rand Corporation to Robert C. Seamans, Jr., Administrator, Energy Research and Development Administration, June 4, 1975.

4. Johnson, L.L., Merrow, E.W., Baer, W.S., and Alexander, A.J., "Alternative Institutional Arrangements for Developing and Commercializing Breeder Reactor Technology", published by Rand Corporation, November 1976.

Chapter VII

Institutions and Commercialization: Possible Patterns

The history of nuclear energy adversities suggests that institutional redirections may be of paramount importance if we are to solve some of the more nagging problems, such as:

- the public acceptance of nuclear energy,
- the political acceptance of fuel recycle;
- the funding of technology development and demonstration; and
- the cost and risk-sharing of commercialization.

The early development of nuclear energy has been characterized primarily by its technology innovation and implementation. The approaching development phase may be recognized by future historians primarily for its institutional innovation and implementation. While institutional innovation may not seem difficult in theory, the ultimate implementation of institutional redirections may be enormously difficult.

Perhaps more than any other group, the Institute for Energy Analysis has grappled with the institutional problems. [1] They note, for example, that the introduction of nuclear energy essentially took nuclear technology as it evolved from the military programs (the nuclear navy in this country and the plutonium-production reactors in other countries) and married it with an existing utility-industry structure. Obviously, consideration was given to safety, reliability, engineering and economics, but primarily within the framework of familiar technology and institutional patterns.

Both the changing issues and the anticipated large future growth of energy demand dictate that more careful thought be given to institutional directions and, possibly, different technology priorities to be more compatible with the required institutional directions. The subsidence in energy growth, and particularly nuclear energy growth, affords an opportunity for such a re-examination. The new directions that might be optimal are not yet obvious, but the needs for those redirections are becoming increasingly apparent. It is the intent here to put emphasis on the needs, not necessarily the solutions. However, examples of redirections will be outlined, primarily to emphasize the fact that the deviations from familiar patterns may, indeed, be significant and, in turn, may involve some modifications of technology directions. The required changes may be particularly significant for the institutions involved in the generation and distribution sector of the energy industry.

Some recapitulation of previous observations will be emphasized as a basis for the subsequent discussion. Those observations are:

1. The traditional nuclear strategy was based on the projection of a continuing high-energy-demand growth as extrapolated from the growth of the 1950-1970 period. Subsequent events now indicate that was an abnormally-high-growth period, but it is argued here that the period 1975-2000 will probably be an abnormally-low-growth period, with higher growth again probable after 2000.

2. The traditional nuclear strategy was also based on economic assumptions that have proven to be inaccurate; but an alternative nuclear strategy making greater use of the thorium cycle is suggested as an expedient for coping with the more difficult economic problems, at least during a technology transitional period.

3. In contrast to the traditional resource and economic concerns, social, political and commercialization concerns are now evolving as much more significant issues.

4. The resolution of these latter concerns will
    almost certainly depend on institutional re-
    directions; including nuclear parks and some
    different definition of government and industry
    roles.

5. Institutional redirections to resolve the more-
    newly-recognized problems may require somewhat
    different emphasis on technology directions--at
    least some alternative technology options should
    be kept open.

6. The development, design and demonstration of
    both the alternative institutional and technology
    directions must be pursued promptly if they are
    to be available for the next energy growth crunch.

Perhaps a common thread through these observations is that
policy planners have become preoccupied with resource and
economic concerns relative to social, political and commercialization concerns, partly because the former have been more
visible, but partly because the latter are more difficult to
define and to resolve.

The following sections of this chapter will explore
potential institutional redirections that may be necessary to
resolve both the technology and the non-technology problems.
Included are discussions of institutional requirements for:

- policy planning;
- research, development and demonstration;
- project management,
- energy generation and distribution; and
- fuel services and supply.

INSTITUTIONAL REQUIREMENTS--POLICY PLANNING

Perhaps one of the most fundamental institutional requirements
is a better national system for selecting appropriate energy
policies and technologies to be pursued. In the past, national

energy directions have been based partly on administration interests, partly on DOE (including AEC and ERDA predecessors) assessments, and partly on congressional judgements--all of which are largely responsive to lobbying by industry, national laboratories and various vocal public-interest groups. Occasionally, input from external panels is utilized, as for example, reports from the Ford Foundation, the Rockefeller Foundation and the National Academies of Science and Engineering. And cost/benefit studies have frequently been used by AEC/ERDA/ DOE to select cost-effective technologies to be pursued.

Recently, DOE has completed a rather exhaustive evaluation of nuclear technologies with emphasis on alternative directions that might offer improved resistance to nuclear-weapons proliferation. The study included evaluations of resource requirements, economics and commercialization problems in the context of promoting proliferation-resistant technologies. Institutional problems were identified, but the report again focused primarily on resource and economic considerations.

Probably the trend toward using external organizations or think-tanks to analyze energy problems, particularly unfamiliar problems, is quite beneficial. This type of independent thinking is likely to identify problems generally overlooked by those closer to policy planning. Nevertheless, strategic planning is so critically important to long-range energy planning, that a more formal national center for problem identification, strategic analyses and policy planning would seem to be appropriate--even to handle assignments to external study groups.

Program analysis in the past has generally focused on technology assessments--primarily the technologies associated with electricity generation in the long range. Until the recent non-proliferation assessment, most of the cost/benefit analyses have been aimed at technologies to achieve more efficient utilization of uranium resources. That goal is obviously an important one in view of projected energy growths in the world for the next fifty years. But, generally, those assessments have not addressed the institutional problems and the technology directions that might be required by alternative institutional directions. The non-proliferation assessment was a promising beginning toward examinations of at least one type of institutional problem. But, even in this case, emphasis was directed primarily toward potential technology fixes for the non-proliferation problem. And, not surprisingly, it was concluded that beyond the once-through fuel cycle there were no technology fixes, i.e., institutional fixes would be required in the long range. But very little attention was given to the details of the institutional redirections and the possible resulting impacts on compatible technology directions.

Still more importantly, the other critical problems of public acceptance, energy independence, commercialization (some attention was, in fact, given to the latter), etc. were beyond the scope of the study. Yet, as has been emphasized repeatedly, solutions to these problems, many of which will require institutional redirections, may be crucially important to energy survival and, ultimately, the survival of society.

INSTITUTIONAL REQUIREMENTS--RESEARCH, DEVELOPMENT AND DEMONSTRATION

If one thing has been proven in the nuclear industry, it is that large research, development and demonstration expenditures for nuclear plants and nuclear fuel cannot be justified by private enterprise. Improvements in energy technology are certainly important to the public and to our national security, but the payback period and the potential profits on nuclear equipment simply do not make private RD&D investments a prudent venture. General Electric attempted to develop the boiling water reactor without government support and probably would not repeat that direction. General Atomic accepted fixed funding for HTGR research, development and demonstration and made enormous investments themselves. It is practically a certainty those investments can never be recovered.

If the nation is to depend on nuclear energy to resolve even partially the growing energy problems, further large RD&D investments must be made. On the basis of previous experience, it is probable that a new generation of advanced LWR plants could require some $1 to $2 billion for RD&D. The RD&D investments for a newer generation of the ACR, such as the HTGR technology, could require some $2 to $4 billion. It is also probable that RD&D for the LMFBR could require an additional $10 to $15 billion. And beyond the reactor RD&D requirements, funding for the fuel cycle might entail another $5 to 10 billion, assuming both the U/Pu and the Th/U-233 fuel cycles are pursued.

The important point is that a very substantial RD&D effort will be required over a period of some 10 to 20 years. While the investments may appear to be high relative to past experience, they may actually be low relative to the financial risks and commercialization costs of technology implementation without strong RD&D programs. It is obvious this magnitude of funding must be forthcoming from government rather than private industry. And, the benefits are obviously for society at large rather than merely for any particular industry.

Assuming, then, that RD&D funding will be a primary function of the government, greater attention must be given to the institutional arrangements to assure that the benefits will be realized by society. Having selected the most deserving technology directions that should be pursued, a mechanism for effective RD&D implementation must be assured. The RD&D activities should benefit from the experience of the supply industries and should be motivated by the ultimate users. The institutional structure that was so successful for naval reactors appears to be an interesting one to consider for future civilian energy development. Mission-oriented design centers operated by industry for the government would presumably satisfy that

criterion. Access to the results of this development and
design work should be available to all interested vendors.
Assignments of personnel from vendor companies to the design
centers might actually expedite that flow of information.

Design centers for light water reactors, high
temperature gas-cooled reactors, fast breeder reactors and
possibly heavy water reactors should, then, be considered for
reactor RD&D including design activities, systems and equipment testing, and dissemination of the technology information
to the vendor industries. The distinction between the naval
reactor laboratories and national laboratories was an important
one; the former being basically design centers and the latter
fundamental R&D centers. That distinction should again be
emphasized relative to the civilian reactor design centers and
the national R&D laboratories. Fundamental research and
development probably should still be assigned to national
laboratories where strong teams have evolved for this kind of
activity. However, specific responsibilities for reactor design,
system design and equipment design should be the responsibility
of the reactor design centers.

Perhaps the most expensive, the most risky and the most
important part of RD&D is the demonstration phase. Traditionally, it has been assumed that the equipment supplier would
gain most from this responsibility and therefore, much of the
financial burden for demonstration plants should fall in this
direction--albeit with some support from government. This traditional philosophy was changed somewhat in the case of the
Clinch River Breeder Reactor project, where the government
accepted a major part of the cost and risk for cost overruns.

At least one particularly frustrating problem has evolved
from the traditional government/industry policies. The supplier
selected for a demonstration program has generally been
expected to contribute to the cost of development and
demonstration, yet that supplier has been required to make all
information available to his competitors, since much of the

funding has come from the government. This only compounds the already-difficult problem of allowing reactor vendors to recover their large front-end investments in a business that typically delays the returns on investments at least ten years, and more likely 15 to 20 years.

Possibly some of these problems might be alleviated, though not necessarily eliminated, by including the responsibilities for the demonstration-plant design and equipment subcontracting with the R&D responsibilities of the design center. Complete access to basic designs, then, could be made equally available to all equipment suppliers. Yet follow-on improvements for commercial plants could still be protected by independent suppliers. Again, though, the intent is not to solve the problems here, but rather to identify directions that have and have not worked in the past and to suggest that more attention must be given to future directions.

INSTITUTIONAL REQUIREMENTS--PROJECT MANAGEMENT

Probably the one most important factor that has made the naval reactors "commercialization" a success has been the NR management principles. As previously noted, NR was basically a representative of the customer, the U.S. Navy. As such, it was responsible for knowing the customer needs, for assuring the appropriate RD&D and for acquiring and installing the equipment. That overall responsibility extends beyond those of the technology developer, the equipment supplier and the architect-engineer individually. Since the user must ultimately bear the responsibility to its stockholders for the successful installation and operation of the equipment, it would seem logical that assurance for that success should lie primarily with the customer organization.

To some extent, that philosophy was followed by the Yankee Electric utilities in the early commercialization of the PWR, though development and design were primarily the responsibilities of the supplier. A greater attempt was made to structure the project management for the Clinch River Breeder Reactor program outside the supplier organization; but in this case, AEC/ERDA/DOE assumed primary responsibility for the development and design. In both cases, it was understandable that development and design management responsibilities were not centralized in a user group, since the utilities were not yet as knowledgeable in these areas as were the other parties. One important move to strengthen the R&D capability of utilities has been the establishment of the Electric Power Research Institute. That enterprise has been very successful, but EPRI was never intended to function as a project management ogranization.

Perhaps the establishment of the Gas-Cooled Reactor Associates in 1978 is the closest approximation to the charter of the NR organization.   While development responsibility for the HTGR is currently shared between GCRA and DOE (as it was in principle with NR), that cooperation seems to be working well. There appears to be a strong effort both on the part of DOE and the utilities to encourage an active and effective GCRA. GCRA activities to date have been predominantly focused on appropriate HTGR R&D programs, but there appears to be strong enthusiasm for initiating a demonstration project. The real measure of the GCRA success will be more apparent as a demonstration program is undertaken.

As with the case of NR, though, the responsibilities of the utility-led project management organizations should go beyond just the management of development and demonstration. That responsibility should include all the requirements for the commercialization of a particular nuclear technology.

INSTITUTIONAL REQUIREMENTS--EQUIPMENT SUPPLY

Assuming RD&D responsibilities are assigned to a nuclear design center and program management responsibilities to the user project-management groups, much of the burden on equipment suppliers is removed. Questions of cost and risk responsibilities for first-of-a-kind and follow-on ventures must, however, be resolved. Accountability must be clearly defined so that unnecessary initial costs and risks can be minimized and so that unavoidab'e penalties are farily distributed.

The mechanism for risk assumption associated with vendor workmanship is clearly a responsibility of the vendor; and contract agreements can adequately handle that potential problem. The risk assumption for systems and component performance related to design and qualification testing by the design center is somewhat more difficult. Some of this risk assumption should be covered by the government to hold them accountable for the adequacy of RD&D funding. The management of the design center must also be held accountable for the quality of the work. This can be accomplished partly by contract renewal negotiations between government and the operating contractors on a regular basis, say, once each two to three years. However, if the operating contractor for design centers is not permitted to share in the profits of the equipment vendors, then the design center cannot be expected to accept financial responsibility for design deficiency.

In the end, then, it probably must be a combination of the government and the users who will share the responsibility of design deficiencies with suppliers having the responsibility of component warranties. This being the case, the utility project management office must have very strong control over management, policies and programs of the design center. That concept will be a difficult one to institute largely because both industrial and national laboratories, with the exception of the naval reactor laboratories, have enjoyed considerable autonomy. Hence, design centers must surrender much of their management control to the utility-led project management office. In turn, the utility project management office must be accountable to the utilities who ultimately must bear at least part of the commercialization risk. Except for the navy, this would be a complete new approach to commercialization.

Agreements by all parties should be defined in the very beginning if appropriate risk assumptions are to be expected in the end.

The mechanism for sharing the risk of potential cost overruns for the first few commercial plants should probably involve a sharing arrangement between the owner utility, a consortium of utilities planning on the first ten plants and the government. Likewise, a formula must be developed for the cost-sharing of the incremental commercialization cost of the first ten plants. These agreements will all undoubtedly require some innovative institutional planning, particularly for the more difficult LMFBR commcialization.

Much has been written on the commercialization of new reactor technologies. A very thorough analysis of alternative institutional arrangements for developing and commercializing breeder reactor technology has been prepared by the Rand Corporation.[2] A gas-cooled reactor commercialization study has also been completed by RAMCO[3] with some discussion of the problems and possible solutions for the commercialization of high-temperature gas-cooled reactor plants. While several potential strategies have been indicated by both these studies and others, much work remains to be done.

INSTITUTIONAL REQUIREMENTS—ENERGY GENERATION AND DISTRIBUTION

Institutional redirections of the electricity-generation industry may be the most significant and the most difficult of all; significant because of the changes necessary to resolve problems of public acceptance, political concerns and commercialization problems; difficult because of the inflexibility of most public-utility-commission constraints. One could add the severe problems of financing new plants, except those problems are hopefully (though not clearly) temporary.

The Institute for Energy Analysis [1] has identified five criteria they suggest to make nuclear energy acceptable in the very long range, viz:

1. physical isolation,
2. strengthened security,
3. professionalization of the nuclear cadre,
4. separation of nuclear generation and distribution, and
5. immortality of institutions responsible for nuclear energy.

Those items are almost self explanatory and will not be redescribed here. They all lead to the concept of dedicated nuclear parks where sensitive nuclear operations as well as the electricity generation are confined to secure reservations. As indicated in a previous chapter, the General Electric Company [4] completed a study in 1975 that found construction and operation advantages for energy parks, but concluded environmental effects from waste heat could establish an upper limit on generating capacity for a single site. The studies looked primarily at parks with about 20 reactors or around 26 GWe of capacity.

The Nuclear Regulatory Commission [5] has also conducted a survey on the feasibility and potential benefits of nuclear energy centers, including power centers, fuel-service centers and combined centers. This survey again concluded it is feasible to construct power-plant centers accommodating up to 20 plants, or fuel-cycle centers, or a combination thereof. It also emphasized, however, that dispersed siting of nuclear facilities continued to be a feasible and practical alternative. One noteworthy observation of the NRC study was that the siting of plutonium-burning reactors in nuclear centers would be particularly attractive.

While it is probably too early to conclude unequivocally that energy parks will be or should be mandatory in the future, certainly the advantages and disadvantages should be carefully weighed both by utility and government planners. In view of the apparent safeguards acceptability of the LWR once-through, low-enrichment-uranium fuel cycle, fuel-security considerations alone do not seem to require isolation of those plants in secure centers, though other considerations may favor it. However, if political concerns continue to discourage the recycle of Pu and U-233 fuels, then the concept of confining bred-fuel utilization to a center containing both the reactors and the fuel service facilities should have considerable appeal.

If the nuclear-center concept is pursued as a long-range policy, implications must be examined relative to:
- desirable reactor technology directions,
- energy transmission problems,
- institutional organization, and
- fuel-cycle technology directions.

Each of these considerations will be explored.

Economics, safety and environmental aspects of nuclear parks may all be important factors in choosing appropriate reactor technology directions for this application. With U_3O_8 at less than \$40 per lb (in 1979 dollars) for the next 10 to 20 years, it will be tempting to base economic evaluations on reactor plants characterized primarily by their low capital cost. Three factors, viz:
- future U_3O_8 prices,
- investment risk (due to accidents), and
- future heat-rejection problems

all suggest that longer-range considerations must not be overlooked.

Typically, reactor plants are expected to have a 30-year life and at least a 10-year construction time. It is likely, then, that nuclear plants yet to be committed will utilize more expensive U_3O_8 for some significant part of their lifetime. Hence, careful consideration must still be given to probable higher U_3O_8 prices and more resource-efficient reactors to use this fuel, in spite of apparent near-term low U_3O_8 prices.

The risk assessment studies of reactor accidents have focused primarily on the risk to the surrounding public. Those studies typically show that the possibility of significant property damage to the public is exceedingly small and, presumably, would be even smaller for reactors located in the more remote nuclear parks. In contrast, though, the financial risk to utility property can be quite substantial, as evidenced by the Three Mile Island accident. Moreover, it is conceivable that the risk to utility property might actually be aggravated somewhat by a trend toward nuclear parks with a relatively large density of reactor plants.

Probably investment-risk concerns will be a more significant consideration for all future nuclear plants. *Possibly* investment-risk concerns will be even a greater consideration for nuclear parks. Reactor technologies to reduce the *probability* of risk by an order of magnitude would, of course, be appealing. Reactor technologies to reduce the *consequences* of risk by an order of magnitude might be even more appealing.

The longer-range environmental impacts of early reactors committed to nuclear parks might also be important. For example, if heat rejection from an energy park containing 20 power plants with a 33% efficiency could be accommodated, the number of plants could be increased to 27 assuming a 40% efficiency. And if some of the heat energy is transported away from the park (as in the CHG/DEG concept, i.e., centralized-heat-generation/decentralized-electricity-generation) still more plants could be accommodated. The important point is that over-commitments to low-efficiency plants in early years could limit the overall number of plants in future years, unless the heat for the latter plants is transported from the center.

It is apparent, then, that desirable directions for reactor technology should, in the long range, include reactors featuring:

- better fuel-resource efficiency,
- less accident risk consequences to the investor, and
- lower waste-heat rejection.

These features should receive special attention in national reactor-development programs.

Energy transmission problems are also affected by the energy-park concept. It has generally been a policy to locate electricity-generation stations within a few miles of service loads to minimize both the transmission investment costs and the transmission losses. Those costs would be increased by energy-park systems, but the General Electric study indicated the economic gains from more efficient construction practices should more than offset the transmission penalties.

An alternative to electricity transmission is heat-energy transmission to decentralized electricity generating stations, i.e., the CHG/DEG concept. This concept could have a number of attractions for the longer-range future, e.g.:

(a) The energy park centralizes all the nuclear engineering functions, thereby allowing:

- on-site nuclear construction shops and personnel,
- sharing of nuclear-operations personnel,
- sharing of nuclear-equipment maintenance personnel,
- on-site regulatory inspectors,
- elimination of off-site fuel handling and transport, and
- on-site safeguards inspectors.

(b) Most of the electricity-generating functions (except for the park load) are relegated to decentralized generating stations allowing:
- most of the electrical engineering functions to be in non-exclusion areas,
- waste heat to be dispersed, and
- some greater cogeneration opportunities at the dispersed generation stations.

(c) Heat storage may be possible, either as sensible heat or chemical energy, at the energy park, thereby allowing nuclear energy to be utilized for peak-loaded electricity as well as base-loaded electricity.

The CHG/DEG concept assumes that heat energy is stored either as sensible heat in a heat-transfer salt or as chemical energy in a reaction, e.g., the reversible reaction:

$$CH_4 + H_2O \rightleftharpoons CO + 3 H_2.$$

As indicated in an earlier chapter, sensible heat storage and transmission might be favored for distances up to about 50 miles with chemical-energy storage and transmission favored for greater distances. Important to either concept, though, is a reactor system that generates high-temperature heat.

Once again, it is emphasized that the CHG/DEG concept is introduced only to illustrate how advanced energy technology options must be kept open to accommodate potentially-desirable institutional redirections. The advanced technology features should include good fuel-resource efficiency, lower accident risk consequences, lower waste-heat rejection and higher-temperature capabilities. With those features, more versatility is available for institutional planning, which in turn should assure a greater acceptance of nuclear energy.

Quite clearly, either the energy-park concept or the CHG/DEG concept could require modifications in the energy generation and distribution institutions, as presently structured. Except for a few of the larger utilities, an energy park would almost certainly involve some shared ownership and operation responsibilities. One possible direction is the formation of independent electricity generation companies with utility consortium ownership. Again, the intent is not to suggest or analyze alternative institutional arrangements for energy parks, but rather to note the need for more long-range planning in this area.

Finally, the nuclear park concept could have profound effects on allowable and desirable fuel-cycle-technology directions. That subject deserves special attention.

INSTITUTIONAL REQUIREMENTS--FUEL CYCLE FACILITIES

The nuclear growth projections discussed in earlier chapters suggested an increasing demand for U_3O_8 beginning shortly after the turn of the century. That increasing demand can bring with it significant price increases for U_3O_8, enhancing the economic incentive for fuel recycle in the U.S. That incentive may materialize even sooner in other parts of the world because of more rapid nuclear growth, a reluctance to expand separative work capability and a desire to insure greater energy independence. For the same reasons, fast breeder reactors may be rationalized even if the capital costs are considerably higher than those of LWR plants.

Fuel recycle will undoubtedly bring with it the traditional concerns of handling Pu and U-233, viz, the industrial hygiene problems, the possible terrorism threats, the potential nuclear-weapons proliferation problems and the public acceptance of a plutonium economy. If the nuclear parks appear to be desirable for confining reactor operations, they would appear to be essential for confining bred-fuel handling. The problems of spent-fuel shipping, fuel reprocessing, fuel refabrication, bred-fuel storage and waste processing all are simplified enormously if they are confined to a dedicated site.

Control of the bred fuels will almost certainly be a national responsibility with international inspectors.

The precedent of government ownership of enrichment facilities, as well as the political sensitivity toward enrichment, will almost guarantee that uranium isotope separation will permanently be a responsibility of the government. And the government moratorium on reprocessing, after an enormous private investment in the Barnwell reprocessing plant, will almost guarantee that industry will never again accept such an investment risk. That background and the current political predispositions would strongly suggest that the entire recycle industry, when it does evolve, will be under the ownership and control of government; possibly with operation under subcontract to industry.

Although reactors would not generally be regarded as a "fuel-cycle facility", it is interesting to note that the basic role of the FBR is likely to be that of a fuel factory, not an electricity factory. With that identification, and with the bred fuel under strict control of the government, the FBR might be compared more accurately to an enrichment plant than a power plant. It is possible, then, that breeder factories also might become part of the government nuclear-fuel domain. Perhaps one significant advantage of this, from an industry point of view, is the elimination of the very substantial FBR commercialization costs that would otherwise fall as a responsibility on industry.

Nevertheless, this apparent trend away from a free-enterprise energy industry must be viewed as a concern for a society whose success can be traced largely to an entreprenurial ethic. For that reason alone, careful thought must be given to the appropriate roles of government and industry in the institutional structure of energy systems.

If, indeed, advanced thermal-spectrum reactors are to become the electricity factories and fast-spectrum breeder-reactors are to become the fuel factories of the future, then again it would seem desirable to maximize the ratio of energy to fuel factories to put primary attention on the end-use industry. This suggests that continuing development of advanced thermal-spectrum reactors is at least as important as development of breeder reactors. That philosophy could become even more important if fusion technology ultimately evolves as a more attractive path toward fuel factories.

One other feature of the fuel-center institutional planning appears to offer interesting opportunities. As has been indicated several times throughout the book, the fuel values assigned to bred fuels, viz., Pu and U-233, can have a profound impact on the economic attractiveness of certain technology directions. The prevailing philosophy has been that the values of bred fuels will seek their own level in an open commercial market. Those values would depend on the willingness of nuclear plant operators to substitute a bred fuel for U-235/U-238 fuel. Moreover, that willingness would depend on the performance of a particular bred fuel relative to other candidates, and the relative cost of handling particular fuels. Hence, the value of a bred fuel would be determined by its neutronic value minus some fabrication penalty.

This free market approach has a number of difficulties, including the following:

1. The unambiguous value for a bred fuel could not be ascertained until a full process industry has been established. The risks of incorrect value assignments could be large in the meantime.

2. In the early period of recycle, the surfeit of bred fuel could also have a significant impact on its market value.

3. If commerce of a particular bred fuel should be limited by national or international controls, the market value of the fuel could deteriorate even more.

4. Under these conditions, an anomolous
   market value for a particular fuel could
   encourage reactor technology in directions
   not desirable for the long range.

If, however, the government were to be the merchants
for bred fuels, this hurdle could probably be eliminated.
Moreover, a government involvement in this enterprise could
encourage appropriate bred-fuel policies relative to non-proliferation interests as well as economic interests. Independent
of bred-fuel pricing policies, it is presumed that plutonium
utilization would be limited to the secure centers. As an
example, then, it might be useful for a bred-fuel bank to assign
low bred-fuel values both to Pu and U-233 initially to
discourage the excess production of Pu and encourage the
purchase and use of U-233. As the recycle industry matured,
the price of U-233 might be increased to encourage the
adoption of still more resource-efficient nuclear reactors.

The example of price adjustment to encourage specific
technology directions is meant only to be illustrative. The
subject of fuel cycle economics in a changing climate of U_3O_8
prices, separative work prices, reprocessing costs, refabrication
costs, working-capital interest rates, inflationary rates,
alternative fuel cycle parameters, etc., is an exceedingly
complicated economic problem. Obviously, the concept of a
bred-fuel-banking institution and its policies is a major study,
but one that should get some attention as part of other reactor
strategy and institutional policy studies.

It is easy to procrastinate on the difficult subject of
institutional planning. It can be argued, for example, that
fuel-recycle facilities and fast-spectrum reactors will not be
required for 20 to 30 years and that the appropriate
institutional organizations can be postponed until a more
urgent necessity arises. This procrastination, however, would
be undesirable for at least two reasons. First, a major fraction

of this country's nuclear technology development program is aimed toward the prevalent strategy. Without revising that strategy, including the institutional planning associated with the strategy, commercialization problems could be enormous without heavy government subsidies or government ownership. Second, and probably even more important, an aggressive planning and initiation of appropriate institutional structures could be very beneficial as a precedent for future planning in other countries. Much has been spoken and written about the merits of secure energy/fuel centers, but without some overt indication of U.S. planning directions, those concepts are only paper studies. To give the subject of secure centers and confined plutonium utilization greater authority, a national or even an international demonstration program of this institutional arrangement could be initiated. By incorporating those features that might be desirable for international implementation, the U.S. could assume a leadership role for world directions.

In summary, considerable progress has already been made on nuclear technology directions, though much remains to be done. Practically no progress has been made on the difficult problems of institutionalization--beyond paper studies. The challenge is great; the time may be short!

REFERENCES

1.   "Economic and Environmental Impacts of a U.S. Nuclear Moratorium, 1985-2010", Insitute for Energy Analysis, MIT Press, 1979.

2.   Johnson, L.L., Merrow, E.W., Baer, W.S., and Alexander, A.J., "Alternative Institutional Arrangements for Developing and Commercializing Breeder Reactor Technology", published by Rand Corporation, November 1976.

3.   "Gas-cooled Reactor Commercialization Study", prepared for the U.S. Energy Research and Development Administration, RAMCO, June 10, 1977.

4.   "Assessment of Energy Parks vs Dispersed Electric Power Generating Facilities", NSF 75-500, Center for Energy Systems, General Electric Company, May 30, 1977.

5.   "Nuclear Energy Center Site Survey-1975", U.S. Regulatory Commission, (NECSS-75), January 1976.

APPENDIXES

Appendix A

The Lore of Logistics

Energy growth can, in principle, be projected by any of several methods. Probably each of the methodologies can be classified in one of the following general categories:
- econometrics,
- end-use integration, and
- trend analysis.

Approaches in the first category usually depend on analyses of interactive economic and energy-demand trends. Those in the second category lean on the integration of supply and demand requirements involving various energy sectors and energy resources. Those in the third category use simple empirical equations based on previous energy-growth history. Hence, the approaches might be described as analytical, integral and empirical, respectively.

The econometric approach has been used in a number of energy studies, only a few of which are referenced here. [1,2,3] The objective of econometric energy studies can be either (a) to minimize the energy cost to the economy by appropriate choices of energy technologies and their resources or (b) to maximize the welfare of the domestic economy by allocating its expenditure budget optimally between energy and other goods and services. Alternatively, the economic penalty to society can be calculated for off-optimum strategies.

A simple and illustrative description of the econometric approach can be found in the Ford Foundation/MITRE report (reference 1). In a "base case", for example, the model used in the report assumed that at *constant* energy cost, domestic energy use and GNP would grow at 3.5% per year for the rest of this century, and at slightly lower rates thereafter. The effects on energy and GNP growth were then examined assuming *increasing* energy costs with alternative price elasticities of demands (negative feedback effects of energy price on energy demand).   Assuming high energy costs and a relatively high price elasticity of 0.5, a domestic energy consumption of about 100 quads is projected for the year 2000.   Assuming a price elasticity of 0.25, a domestic energy consumption of about 150 quads is projected for 2000.

Actually, the purpose of the Ford/MITRE econometrics study was to examine the effect of potentially high energy costs and the delay of nuclear power on the national economy for the next 25 to 50 years. For results of those studies, the interested reader should refer to the reference.

The econometric approach as a tool for projecting energy growth, per se, has the disadvantage that the outcome can be critically sensitive to the economic input assumptions. The more important application of econometrics is the evaluation of alternative national policy directions as measured by the effects on national welfare.

As a tool simply for evaluating energy growth, the end-use-integration approach offers distinct advantages over the econometric approach.   This integral approach has been used by the WAES (Workshop on Alternative Energy Strategies) [4,5] and by the IEA (Institute for Energy Analysis). [6,7]   The approach is a particularly satisfying one because it begins with fundamental assumptions such as population growth, labor productivity, energy efficiency in various energy sectors, etc. The future growth of energy demand, then, is developed for each of the energy sectors (residential, commercial, industrial and transportation) based on the demographic and economic assumptions. In reference 7, data bases are developed and the methods are applied. Low and high projections of 101 and 126 are derived in that study for the domestic energy consumption in the year 2000.

The logistic methodology for growth projection is based very simply on the extension of empirical curves fitting past historical data. The cycle-adjusted-logistic approximation includes the superposition of a cyclic perturbation (again based on empirical data) on a normal logistic growth curve. One important reason for discussing logistic curves further in this appendix is to dispel any possible overconfidence in long-range extrapolations based on this methodology. Clearly, the use of logistic curves to describe and to project growth phenomena offers distinct advantages over linear, polynomial and exponential extrapolation techniques. But the logistic curve like other extrapolation techniques, must be classified basically as an empirical approach to curve fitting. The primary purpose of this appendix is to indicate alternative logistic equations for describing growth and to illustrate the limitations in using these equations.

Logistic curves have been used generally for two purposes in this book, viz:

- for interpolation, and
- for extrapolation

of growth data. As a broad observation, the uncertainty associated with the use of logistic curves to interpolate growth pheonomena in the context of energy growth, is probably less than 5%. The uncertainty arising from the use of logistic curves to extrapolate some 50 years forward may be in the range of 10 to 20%.

Much of the pioneering work on logistic growth curves is attributed to the biologist, Professor Raymond Pearl, [8] who developed and tested his methodology in the 1920s. An excellent discussion of various logistic approaches can be found in the appendix of a recent book by Herman Kahn. [9] Discussion here will focus primarily on the accuracy one might expect from various choices of logistic equations and parameters.

One generalized form of the logistic equation is:

$$n = \frac{N}{(1 + ce^{-\alpha t})^{\beta}} \qquad (1)$$

where n is the growth size at time t. In this equation, N is the asymptotic size at $t = \infty$ and α, β and c are constants that can be chosen to fit past data. In most of Pearl's work, the constant β was chosen to be unity. The accuracy of that approach can be illustrated by some U.S. population-growth projections made by Pearl and Reed [10] in 1924 using an equation similar to (1). In that projection, they forecast a population of about 170 million in 1975, some 50 years forward from their projection date. That forecast was in error by about 20% based on present estimates. Perhaps more significantly, their parameter choices led to an asymptotic population of 197 million; a population that was already exceeded in 1970.

Numerical experiments with the logistic equation show that an increase in the value of the exponent, β, has the effect of increasing the asymptotic value of N for equivalent data fits in the earlier part of the growth curve. Hence, one way to force results showing a longer growth rise is simply to choose $\beta = 2$ or even $\beta = 3$ instead $\beta = 1$. By choosing a data fit for equation (1) with $\beta = 2$, for example, Pearl and Reed probably would have projected the 1975 population more accurately and would have suggested a higher asymptotic population level. That, however, does not necessarily mean this choice of parameters would be infallible for a projection of the next 50 years.

An alternative approach has been used for most of the examples in this book. In this approach, numerical solutions have been found for an exponentially-saturating differential growth curve. The generalized growth curve in this case is described by the differential equation:

$$\frac{1}{n} \frac{dn}{dt} = ae^{-bt^m}, \tag{2}$$

where a, b and m are constants that again can be used for curve-fitting. The curves generated by the difference solutions to this equation can be made to approximate those of equation (1) within 2 or 3%.   Equation (2) has several advantages, however. First, it is intuitively satisfying since it clearly has a pure exponential growth for $bt^m \ll 1$ and an exponential saturation (at least for m = 1) when $bt^m \gg 1$.   Perhaps its greatest advantage is that the choice of the parameter, a, fixes primarily the early growth rate and the parameter, b, controls primarily the terminal growth. The exponent, m, strongly influences the acceleration of the growth saturation. Because of the separation of functions for the various parameters, growth curves can be more easily adjusted to fit data at different time intervals. The disadvantage of equation (2) is that it requires a difference solution that could be tedious without a digital computer. However, the solution is quite simply programmed for a small computer.

Even though equation (2) appears to have an appropriate form to describe a saturating growth curve, it is again emphasized that the primary function of the logistic calculation is to generate an empirical curve that approximates existing growth data. That empirical curve has been useful here for two purposes, viz:

(a) to examine deviations from a normal growth
      curve due to growth irregularities (including
      possible cyclic irregularities), and

(b) to allow some limited extrapolation of growth
      data for purposes of projecting future growth.

In order to be useful for the first of these objectives, though, the uncertainty of the uniform logistic growth must be

significantly smaller than the amplitude of irregularities being
examined. That criterion appears to be satisfied for the energy
growth applications discussed in this book.

Since overall energy-consumption growth depends on
population growth as well as per-capita energy-consumption
growth, it is useful, perhaps, to examine population-growth
projections and energy-growth projections separately.

PROJECTIONS OF POPULATION GROWTH

Statistics are available on U.S. population and population
growth trends for approximately a 200-year period. [11]   These
statistics can be used to select appropriate parameters for
equations (1) and (2) allowing a test of accuracy for the
logistic approximations. Moreover, population projections can,
then, be developed for the next 50 to 70 years.

Figure A-1 illustrates results of population growth
projections using equation (1) with:

$$\beta = 1, \text{ and}$$
$$\beta = 2$$

as well as results using equation (2) with:

$$m = 1,$$
$$m = 1.5, \text{ and}$$
$$m = 2.$$

Data points shown in the figure (including an estimation of
225 million for 1980) suggest that the cases for $\beta = 1$ and $m = 2$
might underpredict the future growth. The case for $m = 1.5$
appears to be the preferred case. The projected populations
for 2025 and 2050 are summarized in table A-1.   Shown also
in that table are the mean and maximum deviations of actual
population data from the logistic estimates.

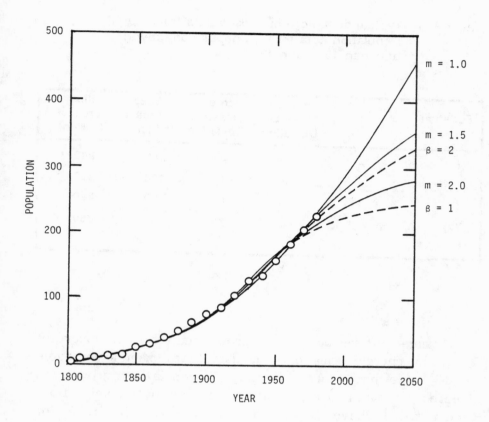

Figure A-1   U.S. population growth as approximated bv various logistic assumptions. (Actual populations shown by points.)

Table A-1   Typical deviations of logistic data from actual
            Population data for U.S. population growth
            between 1800 and 1980

Equation	β	m	Percent Maximum Deviation	Percent Mean Deviation	2025 Forecast (million)	2050 Forecast (million)
(1)	1	–	12	4	240	240
(1)	2	–	8	3.5	300	330
(2)	–	1	9	4.4	380	460
(2)	–	1.5	6	3.0	310	360

Since future population growth can be an important
factor in energy demand, it is useful to examine probable
trends somewhat more carefully.   Population growth is
affected by birth rates, death rates, and net immigration.
Recent data [12] have suggested that the fertility rate for
women (the average number of children borne by a woman)
has decreased from about 3.5 in 1955 to 1.8 in 1975. With no
changes in death rate and with zero immigration, a fertility
rate of 2.1 would be required for a constant population. The
U.S. Bureau of the Census has projected population levels in
2025 based on three assumptions for fertility rates, viz:

Series I (FR = 2.7):   382 million,
Series II (FR = 2.1):   300 million, and
Series III (FR = 1.7):   250 million.

The projected population level is, then, quite sensitive to the
assumed fertility rate.

Figure A-2 illustrates the change in fertility rate for U.S. women from 1900 to 1975 with projections beyond 1975. As further perspective, the fertility rate in 1800 was around 7, with the rate falling uniformly from that point to a rate of about 3.5 in 1900. The rate of decrease actually became more abrupt after 1920, perhaps signaling an advent of improved birth control measures. It is interesting that birth rates began a significant decline before the great depression and showed an increase beginning before the end of World War II. One might speculate that high birth rates tend to be associated with an ebullient economy, in which case one might expect another surge following the current stagnation of our economy. For this reason, and because immigration might be significantly larger than projected (largely due to illegal immigration), it is likely that the population level in 2025 will be significantly above that projected by the Series III forecast--in spite of the apparent low fertility rate at this time. Hence, a population projection around 300 million in 2025 would appear to be reasonable. This would be consistent with the logistic curve using equation (1) and $\beta = 2$, or the curve using equation (2) and $m = 1.5$.

Assuming the use of equation (2) with $m = 1.5$, a U.S. population of 310 million is projected for 2025. If, indeed, a population level of 382 million is reached in 2025, as suggested by the Series I forecast, the logistic growth projection would have been in error by 19%. If the Series III level of 250 million proves to be correct, the logistic error would have been 24%. Deviations as large as these would appear to be very unlikely, hence, the error due to the logistic extrapolation should be within $\pm 20\%$ for a 50-year projection.

Before leaving the topic of population growths, it is interesting to examine world population growths. While the rate of population growth for the U.S. and developing countries appears to be approaching zero, the growth rates in

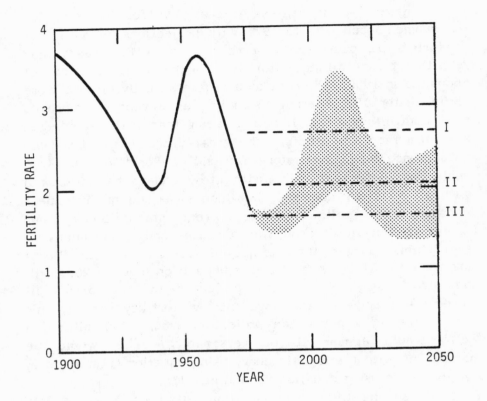

Figure A-2   Approximate fertility-rate variation from 1900 to
             1975. The band from 1975 to 2050 indicates a
             speculative trend. Horizontal lines show three
             cases used by U.S. Census Bureau, identified as
             Series I (2.7), Series II (2.1) and Series III (1.7).

Africa, Latin America and Asia are still quite high. For example, growth rates for the U.S. and Europe are typically around 0.7% per year. In contrast, the growth rate for the world is about 2% per year with the rate for developing countries almost 3% per year. For those growth rates, the population level for the U.S. 50 years hence would be less than 1.5 times the current level, but for the world it would be more than 2.5 times the current level. Both because of the much larger population growth rate and the probable higher per-capita energy growth rate, energy growth in the rest of the world should be much more important than that in the U.S. for the next several decades.

PROJECTIONS OF ENERGY GROWTH

Considerable flexibility is also available for fitting energy growth with logistic equations. Moreover, cyclic perturbations on logistic growth can also be handled, as indicated in Chapters I and II. Again, though, it must be recognized that these approaches are simply empirical. And, projections using logistic or cycle-adjusted-logistic curves make the tacit assumption that energy growth will continue to evolve in the same pattern as that of the previous fitted growth, e.g., the last 150 years in the case of the U.S. total energy consumption. That assumption obviously could be unreliable. It is the intent to examine here the estimated magnitude of uncertainty that might be associated with the various energy projections.

As indicated in Chapter I, deviations of ±20% have regularly occurred, relative to a logistic mean curve, on a 55-year cyclic basis and, indeed, the superposition of such a cycle on the normal logistic growth reduces the deviation between the empirical and actual growth data significantly. But, even with the superposition of the cyclic curve on the logistic base curve, annual energy consumptions deviate, on the average, about ± 5% from the CAL approximation, with some differences as large as 10%. For example, the data point for 1965 falls a little more than 10% below the curve; the point for 1979 some 5% above the curve.

When the uncertainties of future projections are included, the energy forecast assumption some 25 years forward could probably be in error by 10 to 20% and for 50 years forward, perhaps 15 to 30%. The apparently-low CAL projection of 74 quads for U.S. energy consumption in the year 2000 is particularly worrisome. By a somewhat different choice of the logistic parameters, a consumption of 80 quads could have been projected, as shown in figure A-3. Moreover, if one assumes that economic growth is more amenable to government policy directions than was the case in past years, one might argue that cyclic variations could be decreased from the historic amplitude of 20% to, say, 15% or 10%. It is difficult, though, to rationalize a domestic energy consumption of more than 100 quads using the CAL projections. Hence, in Chapter II (figures 2.3 and 2.5), a minimum of 80 quads and a maximum of 100 quads were suggested for limits of U.S. energy consumption in the year 2000. These limits are considerably lower than the values of 101 to 126 quads developed by the IEA [6,7] using the end-use integration methods. However, as indicated in Chapter II, the CAL energy projections for the year 2025 are in general agreement with other sources.

Figure A-4 illustrates how the logistic growth curve, given by equation (2), can be adjusted by the exponent m to force a slower or faster saturation of energy growth. In this figure, projections are carried to the year 2100 simply to show the effect. Using $m = 1.5$, an asymptotic energy consumption of about 270 quads is implied. Assuming an asymptotic U.S. population of 400 million, this suggests an energy consumption per capita of about 23 kwt-yr/yr; compared to about 12 kwt-yr/yr in 1975. A choice of $m = 1.0$

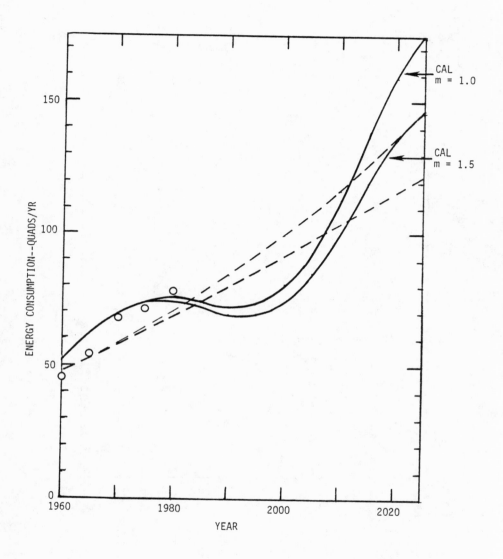

Figure A–3   Cycle–adjusted–logistic projections for U.S. energy
             growth for m = 1.0 and 1.5. Dashed lines
             illustrate unperturbed logistic curves

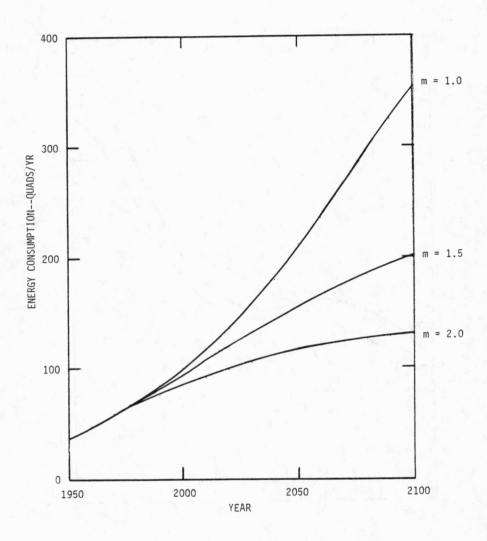

Figure A–4   Three illustrative logistic curves for U.S. energy
consumption

would have led to a per-capita asymptotic energy consumption beyond 50 kwt-yr/yr, while a choice of m = 2.0 would suggest 11 kwt-yr/yr (i.e., no per-capita energy growth). Hence, the logistic equation (2) with m = 1.5 has been used for projections in this book. Approximately equivalent results would have been obtained using equation (1) and β = 2.

While not illustrated here, the logistic equation for world energy growth suggests an asymptotic energy consumption of 8800 quads. With an asymptotic population of 10 billion, this implies a per-capita energy consumption of about 9 kwt-yr/yr, a value about half that for the U.S. Probably this lower per-capita world energy consumption indicates the logistic parameters have been chosen too conservatively for this case. Had parameters been chosen to obtain a worldwide per-capita energy consumption of 20 kwt-yr/yr, growth curves for world total energy consumption would have been somewhat higher than used in Chapter II.

In summary, the empirical logistic curves used in this book suggest that U.S. population is currently about 55% of its asymptotic value and per-capita energy consumption is about 60% of its asymptote. Hence, U.S. total energy consumption is about one-third its ultimate level. In contrast, world population is currently about 45% of its asymptotic value with per-capita energy consumption at about 20% of its asymptote, assuming an ultimate energy-use level of 9 kwt per person. Assuming an ultimate level of 20 kwt per person, the per-capita energy consumption is about 10% of its asymptotic value. Hence, world total energy consumption is only 4 to 9% of its asymptotic value. Once again, it becomes clear that the serious energy problem is one associated with world growth, not just U.S. energy growth.

REFERENCES

1. "Nuclear Power Issues and Choices", Report of the Nuclear Energy Policy Study Group, Ballinger Publishing Company, Cambridge, Mass, 1977.

2. Manne, Alan S., "ETA: A Model for Energy Technology Assessment", The Bell Journal of Economics.

3. Pindyck, Robert S., "The Structure of World Energy Demand", The MIT Press, Cambridge, Mass., 1979.

4. "Energy Global Prospects 1985-2000", Report of the Workshop on Alternative Energy Strategies (WAES), McGraw-Hill, 1977.

5. "Energy Supply-Demand Integration to the Year 2000", (WAES), The MIT Press, 1979.

6. Weinberg, Alvin M., et al, "Economic and Environmental Impacts of a U.S. Nuclear Moratorium, 1985-2010", Institute for Energy Analysis, The MIT Press, 1979.

7. Allen, Edward L., "Energy and Economic Growth in the United States", The MIT Press, Cambridge, Mass, 1979.

8. Pearl, Raymond, "The Biology of Population Growth", Alfred A. Knopf Publishing Co., New York, 1930.

9. Kahn, Herman, "World Economic Development-1979 and Beyond", Morrow Quill Paperbacks, New York, 1979.

10. See Putnam, Palmer C., "Energy in the Future", Van Nostrand Press, New York, 1953.

11. "Historical Statistics of the United States-Colonial Times to 1970", U.S. Department of Commerce, Bureau of the Census, Bicentennial Edition, Parts I and II, September 1975.

12. Westoff, Charles F., "Marriage and Fertility in the Developed Countries", Scientific American, Volume 239, December 1978, pp. 51-57.

Appendix B

Neutronic and Economic Background

It has been emphasized that the overall economics of alternative electricity-generating systems depend both on plant-related costs and fuel costs. With present price levels for uranium fuel, the economic performance of nuclear plants is dominated by plant-related costs. With higher price levels for uranium fuel in the future, fuel cycle costs will become more important. Under the latter conditions, then, nuclear economics will be strongly affected by the efficiency with which nuclear resources are used, i.e., on the neutronic performance of the reactor.

This appendix will review some of the fundamental characteristics of the alternative reactor fuel cycles with special emphasis on the characteristics of U-235, Pu and U-233 fuels used in those fuel cycles. The efficiency with which neutrons are utilized in the fuel cycles can be described by some simple neutron bookkeeping, i.e., neutron balance sheets showing the difference between neutron production (or revenue) and neutron consumption (or expenses), and hence the potential for fuel breeding (or profitability).

The attractiveness of a particular fuel cycle must ultimately be measured by its economic characteristics, i.e., by the fuel-cycle cost. Factors other than just the fuel resource utilization enter into the fuel-cycle cost. Attention must be given, then, to all the components of fuel cycle costs and how these components are related to reactor performance characteristics.

THE ALTERNATIVE FUEL CYCLES

The only fissile material occurring in nature is the isotope U-235. The alternative fissile isotopes, U-233, Pu-239 and Pu-241, can all be generated in nuclear reactors and used in subsequent recycled fuels. These alternative fissile materials do not exist in nature, however, basically because their decay half-lives are relatively short compared to the estimated life-time of the earth's matter.

All fuel cycles, then, must initially use U-235 as the starting fissile material. However, either of two fertile materials, viz: U-238 or Th-232, can be used with the fissile U-235. The two basic fuel cycles for thermal-spectrum reactors are generally characterized by the nuclear reaction involving the fertile material. The two cycles are illustrated in figure B-1.

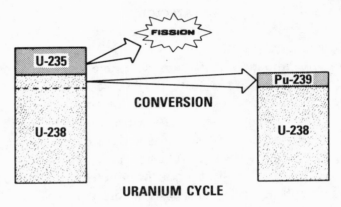

URANIUM CYCLE

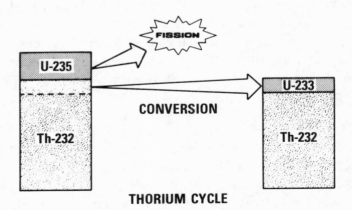

THORIUM CYCLE

Figure B-1   An illustrative comparison of the uranium (LEU) and thorium fuel cycles

In the uranium cycle or, as it is more frequently called, the low-enrichment-uranium (LEU) cycle, approximately 20 to 30 U-238 atoms are used for each one atom of U-235, i.e., the U-235 enrichment is 3 to 5%. In this case, excess neutrons from the fissile process are absorbed in the U-238 fertile material with the following results:

$$n + U\text{-}238 \longrightarrow U\text{-}239$$
$$\downarrow$$
$$Np\text{-}239$$
$$\downarrow$$
$$Pu\text{-}239,$$

where the vertical arrows indicate radioactive decay processes involving the beta decay of U-239 and Np-239.

The other cycle, known as the thorium cycle begins again with U-235 fuel that, in this case, has had most of the naturally occurring U-238 removed, i.e., with high-enrichment-uranium (HEU; about 90% U-235) fuel. Excess neutrons from the fission process are, in this case, absorbed in thorium according to the following reaction:

$$n + Th\text{-}232 \longrightarrow Th\text{-}233$$
$$\downarrow$$
$$Pa\text{-}233$$
$$\downarrow$$
$$U\text{-}233,$$

where, again, the vertical arrows indicate beta decays.

The nuclear reaction chain for the thorium cycle is strikingly similar to that for the uranium cycle. There is, however, one small difference worth mentioning. The Pa-233 loiters longer than Np-239 before decaying to its fissile product.   Because of this procrastination, it not only can absorb an otherwise virile neutron, but it can abort the decay process leading to a fissile nuclide.   Typically, this casualty costs approximately a 0.02 loss in potential conversion ratio for the thorium cycle.   This penalty, however, is small relative to the advantageous neutron-production effectiveness of the bred U-233 fuel compared to that of the Pu-239 fuel.   That advantage, in a thermal-spectrum reactor, can be as large as 0.40 when measured in terms of potential conversion ratio.

With current enrichment technologies, it is significantly more expensive to enrich uranium to 90% than to 3%.   For example, for U_3O_8 at $40 per lb and separative work at $100 per SWU, the cost of U-235 in HEU fuel is about 35% higher than that of the U-235 in LEU fuel.   This higher enrichment cost generally favors the use of the LEU fuel cycle over that of the thorium fuel cycle even though the bred Pu-239 has less favorable neutron-multiplication characteristics (in thermal-spectrum reactors) than does the U-233.   If new enrichment technology should allow a significant reduction in the cost of HEU fuel, the thorium cycle could become more popular.   For example, an enrichment-cost reduction from $100 per SWU to $25 per SWU could decrease the U-235 cost penalty in HEU fuel from 35% to about 15%. It is significant to note that for one advanced thermal-spectrum reactor, i.e., the high-temperature gas-cooled reactor (HTGR), the thorium fuel cycle already is economically advantageous by a significant margin.

A variant of the thorium cycle is one that uses medium-enrichment-uranium (MEU) instead of high-enrichment-uranium (HEU) as the feed fuel. In this case, the fertile fuel contains some U-238 as well as Th-232. Consequently, both U-233 and Pu-239 are bred in this fuel cycle.   The MEU/Th fuel cycle has received considerable attention in recent government studies because of arguments that it avoids the use of weapons-usable fissile materials at the front-end of the fuel cycle and limits to some extent, the amount of weapons-usable materials at the back-end.

However, it is generally recognized that this expedient does not in itself, solve the problem of potential nuclear-weapons proliferation. Moreover, it is significant that the use of the MEU/Th fuel cycle degrades the neutronic advantage of the thorium cycle, thereby putting greater burdens on fuel supply requirements and reactor economics. Because of both these considerations, it is likely that interest in the MEU/Th fuel cycle will be short-lived. In the long range, efforts to control nuclear-weapons proliferation will almost certainly lean more on institutional and political controls than on technological controls.

In both the LEU and the thorium fuel cycles, the traditional concept has been to recycle bred fuel with U-235 makeup in a so-called "self-generated recycle" (SGR) mode of fuel management. In the case of the LEU SGR cycle, the recycled plutonium fuel would constitute approximately 30% of the fissile material under steady-state conditions, with the U-235 in the supplementary LEU fuel accounting for the other 70%. In the thorium cycle, approximately 50% of the self-generated fuel would typically be U-233. Consequently, approximately 30% of the fuel in the LEU SGR cycle would have to be handled in specialized refabrication plants while about 50% of the fuel in the thorium SGR cycle would fall in this category.

Secondary fuel cycles can also use either bred Pu or bred U-233 as the primary fuel or as a makeup fuel. The use of the Pu/U-238 fuel cycle in breeder reactors is one obvious example. Another example is the use of U-233 fuel supplied from an FBR for makeup fuel in an advanced converter reactor. These possibilities, however, must be regarded as longer-range strategies.

In the nearer term, the roles of Pu and U-233 are more uncertain. Both economic and political considerations make the SGR use of Pu in LWR plants questionable. Consequently, unless a very substantial increase occurs in U_3O_8 and/or separative-work prices, an alternative strategy for using plutonium will have to be sought.

The use of the thorium cycle with HEU fuel feed has also been subject to some criticism, primarily because of concerns about the large-scale use of weapons-grade, high-enrichment uranium. If and when it does become advantageous to utilize bred fuels, it would be very desirable to move in the direction of the long-range strategy where plutonium will be used primarily in fast-spectrum reactors and U-233 (instead of U-235) in thermal-spectrum reactors.

Presumably the attractiveness of any particular strategy must depend on its economic merits, which in turn must depend partly on the "neutron efficiency" of the strategy. The neutron efficiency depends both on the efficiency of neutron production and the efficiency of neutron consumption. In the next section, attention will be focused on the neutron-production effectiveness of the various fissile nuclides.

VIRILITY INDICES FOR FISSILE NUCLIDES

Each of the various fuel cycles has its own neutronic character-profile that depends largely on the neutron fecundity of the nuclear fuels involved. The number of neutrons available for the conversion of fertile material to fuel is maximized by:

- the use of fissile nuclides having the greatest
  neutron-production capability, and
- the selection of a reactor design that minimizes
  neutron losses to non-productive purposes.

As will be seen, the neutron-production effectiveness differs significantly for the different fissile nuclides.

The neutron-production effectiveness, η, for a fissile nuclide is a measure of its neutronic virility. It is characterized by the net number of neutrons produced from the fission process per neutron absorbed *in the fissile material*. Neutrons absorbed in the fissile material include both gainfully-absorbed neutrons and parasitically-absorbed neutrons. Gainfully-absorbed neutrons produce nuclear fissions in the fissile mat-terial. Parasitically-absorbed neutrons manage only to be captured by the fissile material with no further consequence. The ratio of parasitic capture to fission absorptions for a reactor fuel is an important neutronic parameter generally referred to as the capture-to-fission ratio. The capture-to-fission ratio is quite different for the different fissile nuclides, and is also very sensitive to the energy spectrum of the reactor neutrons.

If the capture-to-fission ratio is large, then the neutron-production effectiveness tends to be small. In general, the neutron-effectiveness tends to be more favorable for neutrons that have been thermalized in a thermal-spectrum reactor, and tends to be less favorable for neutrons that are absorbed when they are still in the epithermal energy range.

Assuming approximately 30% of the neutron absorptions in fuel occur from epithermal-energy neutrons and 70% from thermal-energy neutrons, then the neutron-production ef-fectiveness for each of the various fissile materials is as shown in table B-1. The neutron productivity for U-233 appears to be best and that for Pu-239 poorest for the neutron-spectrum characteristics associated with typical thermal-spectrum reactors. In a fast-reactor spectrum, the Pu-239 would excel.

The overall average neutron productivity for a mix of reactor fuels depends on the fraction of the various fissile nuclides normally present in a particular fuel cycle. For example, the normally-good $\bar{\eta}$ for U-233 fuel produced in the thorium cycle can be largely offset by the poorer $\bar{\eta}$ of U-235 if large amounts of U-235 makeup fuel are required to compensate fuel depletion in the fuel cycle.

Table B-1   Comparison of neutron-production effectiveness
             for alternative fuels

	U-233	U-235	Pu-239
n_{th}	2.29	2.07	1.87
n_{epi}	2.14	1.62	1.76
$\bar{n}$	2.24	1.93	1.84

(n = neutrons produced in fission/neutrons absorbed in
   fissile material)

NEUTRON BOOKKEEPING

The neutron-production effectiveness, then, determines the
neutron supply or the revenue in the neutron balance
sheet. The debit entries in the balance sheet include all the
neutron losses that can occur as a result of neutron leakage
and absorption processes.   The potential loss mechanisms
include:

- neutron leakage from the core,
- absorption in moderator and structural materials,
- absorption in heavy-metal posions,
- absorption in fission-product poisons, and
- absorption in control poisons.

Typically, the neutron absorptions in moderator and structural
materials per neutron absorbed in fissile material is 0.03 to
0.10, depending on the reactor type. The fractional absorptions
in heavy-metal poisons (particularly U-236 and Pa-233) tends to
be in the range of 0.02 and 0.10.   Fractional absorptions in

fission-product poisons usually constitute the largest debit entry, being in the range of 0.10 to 0.20, the exact amount depending on how long the fuel (and fission product poisons) is held in the reactor, i.e., the fractional fuel burnup.   Losses in control poisons can also be as large as 0.10.

It must also be recalled that one neutron absorption in fissile material is necessary to sustain the chain reaction. Hence, the total neutron losses are 1+L where L is the sum of the non-productive losses previously indicated.

The difference between the source neutrons and lost neutrons, then, is a measure of the "profit" neutrons in the balance sheet.   These excess neutrons are available for absorption in the fertile material to allow conversion into a new fissile nuclide.   Since all neutron events in the balance sheet are measured relative to neutrons absorbed in the fuel, the excess neutron fraction per neutron absorbed in fuel is simply a measure of the conversion ratio, i.e., the fertile material converted to fuel per fissile fuel burned. The conversion ratio for a fuel cycle is an important measure of the neutron (and, therefore, fuel resource) utilization of a reactor. The net fuel required to sustain operation of a reactor is the difference between the fuel burned and the fuel produced, which is proportional to (1 - CR).   For example, the net fuel requirements for a reactor with a CR = 0.9 would be half that for one with a CR = 0.8, and only 10% that of a non-converting or burner reactor.

Some typical fuel mixes and resulting neutron balances for illustrative LWR and HTGR fuel cycles are shown in table B-2. The columns indicate the neutron balances (and potential conversion ratios) for the SGR LEU cycle, the SGR Th cycle, and the U-233-fed Th cycle for the two reactor types. It is emphasized that these data are intended only to be illustrative.

Several significant conclusions are apparent from the table as follows:

1. The conversion ratios achievable with the HTGR

Table B–2   Examples of neutron balances for some alternative fuel cycles

	LWR			HTGR	
	LEU SGR	Th SGR	U–233 Th	Th SGR	U–233 Th
Bred–fuel abs/ U–235 abs	0.5	0.6	0.9	0.7	0.9
Neut prod/abs	1.95	2.12	2.21	2.14	2.21
Net losses/abs	0.37	0.40	0.36	0.38	0.31
Neut abs in fuel/abs	1.00	1.00	1.00	1.00	1.00
Conversion/abs	.58	.62	.85	.76	.90

tend to be slightly higher than those with
the LWR;

2. The thorium cycle allows somewhat higher
conversion ratios than the LEU cycle; and

3. A conversion ratio close to unity can be
achieved when U-233 is used as the initial
makeup fuel.

The U-233/Th fuel cycle is particularly interesting. In the long range when U-233 might be available as a by-product, either from FBR plants or fusion breeders, the balance sheets indicate that very little makeup fuel would be required to sustain reactors using this fuel cycle. In fact, relative to the LEU *once-through* fuel cycle currently used by the nuclear industry, a reactor having a conversion ratio of 0.9 would require only about 10% as much makeup fuel.

The Bettis Laboratory [1] has shown that a conversion ratio *greater* than unity can actually be achieved with a carefully-designed LWR plant under some constraints on the fuel cycle. In their light-water breeder-reactor design, the normal posion control rods are replaced with a novel control system that depends on changes in the core geometry. In addition, the specific fuel inventory is chosen about twice that normally used in the LWR, and a fuel burn-up life about half the usual fuel lifetime is specified. With these compromises in reactor design, fuel loading and fuel burn-up, they demonstrate that fuel breeding is feasible with the thorium fuel cycle, even in an LWR plant. While these design and fuel cycle choices would involve substantial compromises in the economics of electricity generation, it would appear that some of the design features of such a reactor might still be utilized for a near-breeder LWR. The possibility of near-breeding, or perhaps breeding, seems to be even more easily achievable with the HTGR plant.

It is probably worthwhile to emphasize once again that breeder reactors, while interesting for the very long range, may not be critically essential or even economically desirable in, perhaps, the next 50 years. The more important resource-related objective of nuclear reactors is to minimize potential energy-generation cost penalties that could arise from increasing U_3O_8 prices. Hence, primary attention should be focused on the economics of alternative reactors and fuel cycles (including breeding cycles) in the face of plausible increases in U_3O_8 prices.

ENERGY ECONOMICS

As was indicated in Chapters III and IV, approximately 65% of the generating cost for nuclear plants is associated with the fixed capital charges, 5% with operating and maintenance costs and 30% with the fuel cycle costs. Moreover, some 5 to 15% of the generating cost can be ascribed directly to uranium-resource expenses. With an LWR plant using the LEU once-through fuel cycle, a generating cost increase of 25 to 30% would be expected, then, for a U_3O_8 cost increase from, say, $40 to $120 per lb. With fuel recycle, this penalty might be reduced to, perhaps, 20%; and with a more resource-efficient fuel cycle, the penalty might be reduced to around 10%.

It is important to recognize, however, that the fuel-cycle-cost contribution associated with the purchase and consumption of U_3O_8 is but one part of the overall cost of generating electricity. One must, then, assure that a particular fuel cycle or reactor concept designed to minimize the U_3O_8 cost contribution does not introduce other economic penalties that might more than compensate the benefit. Hence, all aspects of reactor economics must be carefully weighed in evaluating the relative merits of reactors and fuel cycles.

Although the emphasis in this book has been directed toward fuel-cycle economics, some discussion of plant-related costs has been included both for perspective and completeness. In the discussions of both plant and fuel-cycle economics, the intent was to illustrate principles, not to prescribe a methodology for economic evaluations. Consequently, the further discussion of plant and fuel-cycle economics is intended simply to illustrate how various reactor and fuel-cycle performance characteristics affect economics.

THE PLANT COST CONTRIBUTION

The plant capital cost contributes to the overall electricity-generating cost in three ways, viz, through:

- interest charges against the plant-cost investment,
- amortization of the investment, and
- property taxes and insurance.

The combination of interest, amortization and tax expenses is usually lumped into an annual charge rate, r, against the plant. The product of this annual charge rate and the total cost of the plant, C_p, is simply the plant-related annual cost of owning the plant. Dividing this annual cost by the killowatt-hours per year of electricity generated by the plant yields the component of generating cost in, e.g., mills/kwh, that must be assigned to the plant itself. Obviously, the effective fraction of time the plant operates during the year, i.e., the load factor, is important in assuring economic utilization of the plant.

The annual charge rate, r, is strongly affected by the inflation rate of the economy, i.e., during high-inflation periods, the annual rate might be as large as 18 to 20%; during a zero-inflation period it is typically 10%. One might expect, then, that the fixed annual charges against the plant would cause plant costs to be more important relative to fuel-cycle costs in periods of high interest rates. Actually, it has been shown [2] that inflation rates tend to affect the fuel cycle in much the same way as they affect the interest charges on the plant capital cost; assuming the U_3O_8 and separative work prices change in a consistent way with inflation. This correspondence principle between plant and fuel cycle costs results if the levelized fuel-cycle cost over the operating life of the plant reflects the same inflationary factors as those affecting the plant cost. As a consequence of this correspondence principle, the levelized fuel-cycle costs can be described conveniently as a fraction of the annual fixed charges for the plant. One advantage of that approach is that inflationary effects can be removed. This assumes, however, that the inflation rate before and during construction of the plant are similar to those affecting the fuel cycle economics during operation.

THE FUEL CYCLE COST CONTRIBUTION

Fuel-cycle costs for nuclear reactors have two basic contributing components, viz:
- the fuel-resource-supply cost, or fuel-supply cost, and
- the fuel-service cost, or handling cost.

Generally, the fuel-supply cost, which includes the supply of both U_3O_8 and separative work, accounts for some 80% of the total fuel-cycle cost for once-through fuel cycles at current U_3O_8 prices; but could account for almost 90% of the total for U_3O_8 prices at $120 per lb. In contrast, for the more efficient fuel-recycle mode of fuel management, the fuel-supply cost fraction decreases to about 60% of the total fuel-cycle cost, both because of the lower resource-supply requirement itself and the higher handling cost. It is a strange coincidence, though, that the increased handling cost for recycle approximately compensates the decreased fuel-supply cost associated with the recycle of residual fuels in LWR plants--at least for contemporary economic conditions. However, if the price of U_3O_8 should rise by a factor of two or three times, then significant cost differences would become apparent.

The fuel-supply cost portion of the fuel-cycle cost is affected by the reactor and fuel-cycle performance characteristics through the fuel depletion and working-capital costs. Both of these contributions are affected by U_3O_8 resource and separative-work prices. Figure B-2 shows the fuel-cycle cost for the LWR SGR fuel cycle with the fuel-supply cost first divided according to the fuel-depletion and fuel-working-capital components and secondly, divided

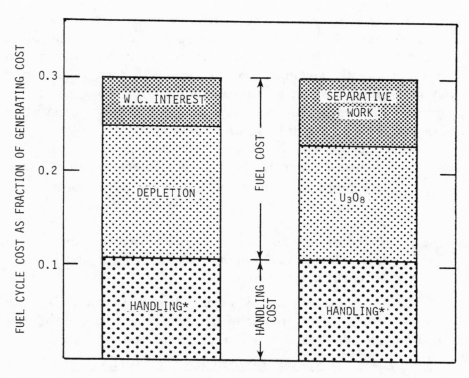

*including working-capital charges against handling costs.

Figure B-2 Alternative fuel cost definitions

according to the U_3O_8 and separative-work price contributions. The first way of illustrating the fuel-supply cost is convenient as a means for examining the effects of reactor conversion ratio and specific inventory on the fuel-cycle cost. The second way of illustrating the fuel-supply cost is convenient for purposes of identifying how changes in U_3O_8 or separative work prices might modify the fuel-cycle cost.

The depletion cost and working-capital cost are measures of how efficiently the reactor utilizes the fuel, i.e.:

1. the depletion cost is related to the net amount of fuel consumed per unit of energy produced, and
2. the working-capital cost is related to the amount and value of fuel required in the reactor and out-of-reactor system.

The depletion cost is measured quite simply by the change in fuel value from the beginning to the end of the burn-up cycle, divided by the energy generated. Hence,

$$c_d = \frac{(\text{initial fuel value} - \text{final fuel value})}{\text{energy}}.$$

The numerator involves both the initial and final quantities of fuel and their values. Hence, if more than one type of fuel is involved in the fuel cycle, the total of the initial and the final fuel values must each be summed. The denominator represents the energy, usually measured in kilowatt-hours, generated throughout the fuel-residence time in the reactor.

Where the initial and final fuel species are the same, the relationship between the fuel depletion cost and reactor-conversion ratio is particularly simple. As fuel is consumed during reactor operation, new fuel is bred--the amount being equal to CR x (total fuel burned). The net fuel burnup is, then,

$$\text{net fuel burned} = \text{total fuel burned} - \text{fuel bred}$$
$$= (1 - CR) \times (\text{total fuel burned}).$$

Hence, the fuel depletion cost, which is related to the net fuel burned, is directly proportional to $(1 - CR)$. In general, then, the depletion cost is large when the conversion ratio is small, and approaches zero as the conversion ratio approches unity.

The working-capital cost reflects the cost of money to cover the value of fuel held in the system. Hence,

$$c_{wc} = \frac{(\text{average fuel value}) \times (\text{interest rate}) \times (\text{time in system})}{\text{energy}} .$$

As was the case for the depletion cost, the fuel value depends both on the quantity of fuel involved and the unit value of the fuel. Consequently, a large specific inventory, i.e., fuel inventory per kwe rating of the reactor, results in a large working-capital cost against the fuel. Moreover, since HEU and U-233 fuels tend to have higher unit values than LEU and Pu fuels, the working-capital cost for fuels in the thorium cycle tend to be higher than those for fuels in the LEU cycle.

The fuel-service or fuel-handling cost includes:

- the fabrication cost for the fresh fuel,
- the refabrication cost for the recycled fuel,
- the spent fuel shipping cost,
- the spent fuel reprocessing cost, and
- the waste handling cost.

The total fuel-handling cost is usually identified by the product of the heavy-metal content of fuel processed each year and the sum of the unit process or service costs. The unit costs are generally expressed by the cost per kg of heavy metal processed for each of the components. The handling-cost portion of the fuel-cycle cost is, then, the sum of the annual costs divided by the energy generated during the year, i.e.:

$$c_h = \frac{C_{fab} + C_{refab} + C_{shpg} + C_{repro} + C_{waste}}{energy}.$$

    The numerator tends to be dominated by the costs of fabrication (and refabrication) and reprocessing (or permanent storage in the case of a once-through cycle).

    In this illustrative discussion of fuel cycle costs, the time-weighted effects of non-uniform expenses has been ignored. In practice, utility expenses tend to be heavier in the early years, largely because the initial fuel inventory and handling costs are substantial relative to the follow-on annual costs for these items. Yet the revenues from the sale of energy are generated more uniformly during the operation of the reactor. To define fuel cycle costs more accurately, it is necessary to calculate cash-flows for expenses and revenues on a properly-discounted basis. Hence, fuel cycle costs are generally expressed as levelized costs with appropriate present-worth values for the expenses and revenues. Typically, the levelized costs are 10% to 20% greater than equilibrium costs that would be obtained from the equations previously described. Data used in Chapters IV and V reflect levelized fuel-cycle costs.

REFERENCES

1. "Design of the Shippingport Light Water Breeder Reactor", LWBR Development Program, Westinghouse Electric, WAPD-TM-1208, January 1979.

2. Stauffer, T.R., Palmer, R.S., and Wyckoff, H.L., "Breeder Reactor Economics", prepared for Breeder Reactor Corporation, July 1975.

ABBREVIATIONS AND ACRONYMS

Institutions

AEC	Atomic Energy Commission
AGNS	Allied General Nuclear Services
AQCR	Air Quality Control Regions
CONAES	Committee on Nuclear and Alternative Energy Studies
DOE	Department of Energy
EEI	Edison Electric Institute
EIA	Energy Information Administration (Department of Energy)
EPRI	Electric Power Research Institute
ERDA	Energy Research & Development Administration
HEC & LEC	High-Energy Consuming & Low-Energy Consuming (Countries)
IAEA	International Atomic Energy Agency
IEA	Institute of Energy Analysis
KAPL	Knolls Atomic Power Laboratory
NR	Naval Reactors
NRC	Nuclear Regulatory Commission
OECD	Organization for Economic Cooperation and Development
OTA	Office of Technology Assessment
WAES	Workshop on Alternative Energy Strategies

Reactors

ACR	Advanced Converter Reactor
BWR	Boiling Water Reactor
CRBR	Clinch River Breeder Reactor
FBR	Fast Breeder Reactor
HTGR	High Temperature Gas-Cooled Reactor
HWR	Heavy Water Reactor
LMFBR	Liquid Metal Fast Breeder Reactor
LWR	Light Water Reactor

LWBR	Light Water Breeder Reactor
PWR	Pressurized Water Reactor

Units

kw	killowatt(s)
kwh	killowatt-hour(s)
kw-yr	killowatt(s)-yr(s)
kwe	killowatt(s) (electric)
kwt	killowatt(s) (thermal)
MWe	megawatt(s) (electric) (= 1000 KWe)
GWe	gigawatt(s) (electric) (= 1000 MWe)
bbl	barrels
BTU	British thermal unit
lb	pound
quad	quadrillion (10^{15}) British thermal units
ST	short tons (2000 lbs)
SWU	separative work unit

Miscellaneous

AIST	Advanced Isotope Separation Technology
CAL	cycle-adjusted-logistic
CHG/DEG	centralized-heat-generation/decentralized-electricity-generation
CR	conversion ratio
GNP	gross national product
INFCE	International Nuclear Fuel Cycle Evaluation
IURE	International Uranium Resource Evaluation
NASAP	Nonproliferation Alternative Systems Assessment Program
NURE	National Uranium Resource Evaluation
O&M	Operation & Maintenance (cost)
PRDP	Power Reactor Demonstration Program
RD&D	Research, Development & Demonstration
SGR	self-generated recycle
TCP	thermochemical pipeline

LEU   low-enrichment uranium
MEU   medium-enrichment uranium
HEU   high-enrichment uranium

INDEX

ABOUT THE AUTHOR

B.S.   (Physics/Chemistry)   Kent State University, 1938
M.S.   (Physics)   Ohio State University, 1939
Ph.D. (Physics)   Ohio State University, 1947

Dr. Stewart is the owner and president of NUTEVCO, a nuclear technology evaluations company located in Sorrento Valley, San Diego, California. NUTEVCO has been doing energy technology consulting for both industry and government.

Until April 1977, Dr. Stewart was Vice President of HTGR Engineering at General Atomic Company. He joined GA in 1959 as the leader of a physics group responsible for the nuclear design of the Peach Bottom HTGR plant. He subsequently held positions as Chairman of the Nuclear Analysis and Reactor Physics Department, Assistant Director of the Laboratory, and Vice President of HTGR Fuel prior to becoming the Vice President of Engineering.

Dr. Stewart was with the Knolls Atomic Power Laboratory of General Electric Company from 1947 to 1959. While there, he was initially involved in a sodium-cooled breeder reactor program and subsequently, served as manager of reactor physics for an intermediate-energy, sodium-cooled submarine reactor and manager of reactor physics for a pressurized-water surface ship reactor.

In his 30 years of experience in nuclear energy, he has worked on liquid-metal-cooled reactors, pressurized-water reactors and gas-cooled reactors. He has written many papers for national and international journals and has contributed to several books.